Lipizzaner

The Story of the Horses of Lipica
Commemorating the 400th Anniversary of the Lipizzaner

CONTROL DATA ARTS
a service of
CONTROL DATA CORPORATION

St. Paul, Minnesota
1981

Lipizzaner: The Story of the Horses of Lipica by Dr. Milan Dolenc
Translators: Žarko Harvat and Susan Ann Pechy
Editor: Marjan Krušič
Consultants: Prof. Dr. Jože Jurkovič, Prof. Dr. Stjepan Romić, Marica Hočevar
Color photography: Emilio Rossi
Cartographer: Stanislav Valič
Designer: Borut Kovše
Editor, American edition: Pamela Espeland
Designers, American edition: Ned Skubic and Evans & Smith Graphic Design
Photograph on page 19 courtesy Walt Disney Productions.

Library of Congress Cataloging in Publication Data
Dolenc, Milan.
    Lipizzaner, the story of the horses of Lipica.

    Translation of Lipica.
    Bibliography: p. 166
    1. Lipizzaner horse. 2. Horse-breeding—Yugoslavia—
Lipica, Slovenia. 3. Lipica, Slovenia. I. Title.
SF293.L5D6413   636.1'3   80-28576
ISBN 0-89893-172-X

Copyright © 1981
Mladinska Knjiga, Ljubljana, Yugoslavia
All Rights Reserved
Printed and bound in Yugoslavia

Lipizzaner

Contents

vii	Preface .... Pamela Espeland
ix	Notes on the 400th Anniversary of the Lipica Stables .... Mitja Ribičič
xi	On Karst .... Ciril Zlobec
xiii	Lipica and the Lipizzaner .... Andrej Franetič
	The Lipizzaner and the Art of Riding
	The Schooling of the Lipizzaner
	Lipica Today
1	*A History of the Lipizzaner* .... Milan Dolenc
	Beginnings and Growth
	The Influence of the Habsburgs
	Hard Times: The Wars with France
	Renovation and Renewal
	More Hard Times: The World Wars
	The Post-War Years and the Struggle for Survival
27	*The Lipizzaner Outside of Lipica* ..... Milan Dolenc
	In Slovenia
	In Croatia
	In Serbia and Vojvodina
	In Bosnia and Hercegovina
	In Macedonia
	In Hungary
	In Czechoslovakia
	In Romania
	In Poland
	In Italy
	In Austria
	At the Spanish Riding School in Vienna

43 *The Development and Breeding
   of the Lipizzaner* .... Milan Dolenc
      The Karst Horse
      The Spanish Horse
      The Arabian Horse
      The Formation of the Various
        Lipizzaner Lines
      The Lipizzaner's Appearance

59 *Lipica and Its Surroundings* .... Milan Dolenc
      The Buildings of Lipica
      The Countryside
      The Towns and Villages
      The Estate

73 The Plates

159 The Plate Captions

168 Bibliography

Preface

The country called Yugoslavia was born in 1918, at the end of World War I, when it and a number of other new states — including Czechoslovakia, Poland, and Austria — were carved out of the ruins of the massive Austro-Hungarian Empire. In 1921 it was declared a constitutional monarchy; in 1945, following yet another devastating World War and the collapse of Nazi Germany, it was proclaimed a Federal People's Republic. For a time it was allied with the Soviet bloc, but a disagreement over the course Yugoslavia would take in the future occasioned a major split in 1948. At that point Yugoslavia struck out on its own and moved steadily in the direction of socialism, developing a new form of self-government and gradually emerging as a strong non-allied nation. A new constitution in 1963 established it as a distinct Socialist Federal Republic incorporating the individual republics of Slovenia, Croatia, Bosnia and Hercegovina, Montenegro, Macedonia, and Serbia, in addition to the autonomous regions of Vojvodina and Kosovo. Today Yugoslavia borders seven nations — Italy, Austria, Hungary, Romania, Bulgaria, Greece, and Albania — and has a population of over twenty million.

     A land of mountain ridges, crisscrossing rivers, a stunning Adriatic Sea coastline, and brilliant blue skies, Yugoslavia is a mixture of ancient history and customs, ethnic diversity, rich folk traditions, and contemporary cultural values. It is a truly multinational society which bears the stamp of Slavic, Eastern, and Western influences alike. Its people include primarily Serbs, Croats, Slovenes, Bosnian Moslems, and Macedonians, as well as Albanians, Hungarians, Turks, Romanians, Czechs, Italians, Greeks, and Jews — all of whom have found a place in the national framework. Several languages are spoken and three major religions espoused: Roman Catholicism, Muslim, and Eastern Orthodox.

     In Slovenia, at the northwestern tip of Yugoslavia, there is a region called the Karst that is known for its underground caves and harsh climate. Tucked away on a green oasis in the Karst is the estate of Lipica, which covers approximately 770 acres and is surrounded by a stone wall. It was here, over four hundred years ago, that Lipizzaner horses were first bred. Today, in spite of centuries of wars, politican upheavals, hardship, and change, Lipica survives virtually intact, a Renaissance jewel which still houses the main stables of the Lipizzaner breed. Troupes of these famous white stallions tour the world, delighting audiences everywhere, and tourists flock to Lipica to see the horses executing the graceful maneuvers of the *haute école* and frolicking on their rocky native ground.

    Known for their proud bearing, elegant gait, superior intelligence, and strength, the Lipizzaner has had a long and exciting history. In words and pictures, this book tells it all, from the beginning to the present day.

Notes on the 400th Anniversary of the Lipica Stables

It is no accident that the breeding of Lipizzaners in Slovenia has gone on for four full centuries. Countless horse enthusiasts, both known and unknown, have worked to preserve this noble line on Slovene soil. In 1959 President Tito himself prevented the closing of the stables, giving official approval to what had long been an honored Karst tradition.

So the four hundredth anniversary of the Lipica stables is a fitting occasion for a more than ordinary celebration. For thousands of years, horses have been indispensable and faithful companions to humankind — in war and in peace, in work and in play. Almost since the beginning of the breed, the Lipizzaner has had a reputation for being one of the finest horses in the world. It deserves to be cherished and maintained, protected and praised.

As we anticipate the next several decades, we are faced with a number of responsibilities. We must continue to keep accurate records of the Lipizzaner's history. We must consider new ways of improving and refining the breed. We must make the equestrian sport available to more people — young and old, rich and poor — and strive to increase their appreciation of the Lipizzaner's special qualities. But most of all we must express our gratitude to the dedicated people of Lipica, who have cared for these magnificent horses for so many years and will undoubtedly continue to do so for many more to come.

On Karst

To a Slovene linguist, the word *karst* is rich with meaning. There is, first of all, the Karst itself — a region surrounded by others which only emphasize its uniqueness; a land of unusual geographical and geological phenomena, including a characteristic rocky landscape, incredible subterranean rivers and caves, gently undulating vistas, and rare patches of fertile soil. Any fields worth cultivating are ringed by small villages, little sentries that guard whatever acres have been seized from the neighboring wasteland and made fruitful. The Karst is so distinctive that its name has been given to similar areas around the world — areas that could easily have remained unfertile and uninhabited yet have been pulled and prodded into shape by people who were not afraid of hard work.

    The word *karst* has several etymological relatives, including *krasen,* meaning beautiful, and *krasota,* meaning splendor. An archaic Slavic term, it carries a number of poetic and romantic connotations for every Slovene. It symbolizes both the land and its people. It stands for the peculiar architecture of the region's villages with their doors and windows framed by Karst limestone, their gardens, and their ornate fountains. It embodies the piles of stones that dot the countryside, dug up laboriously from the soil and arranged into landmarks or walls that border nearly everything — meadows, village lanes, estates, farms, and courtyards. One can never escape the stones, and they continually remind the Slovene people of how arduous life in their chosen land can be.

    Economic conditions in Yugoslavia have vastly improved following World War II, and mere survival is no longer an issue for most Slovenes. Today they are able to view the Karst from a new perspective, focusing on its beauty rather than its barrenness. And there is much beauty to be seen: vast treeless meadows, thick dark green pine woods, lush vineyards, echoing caverns, winding rivers, a sky the color of the Mediterranean, and even the air itself which, in the words of the great Karst poet Srečko Kosovel, is "transparent as the wings of a dragonfly."

    In the midst of all this beauty lies the estate of Lipica, home of the Lipizzaners. It is set like a gem in the middle of a typical Karst landscape among gnarled oak trees, low shrubbery, rocks, and small dales. For centuries it has been tended and molded, cared for and preserved. Today the estate and its horses are known all over the world — and together they continue to give the word *karst* new meaning.

Lipica and the Lipizzaner

Ever since the days of the Roman Empire, Karst horses have been highly valued. Lipizzaners especially have been prized by rulers and aristocrats, who in centuries past harnessed them to splendid carriages, adorned them with elegant trappings, and drove them for pleasure and parade. The commanders of many armies rode on their backs, convinced that their own personalities and stature were reflected in the strength and grace of their superb white stallions.

Humankind has used the horse in varying capacities for thousands of years. Once tamed, it was a faithful servant and companion in times of war and a willing beast of burden in times of peace. But as mechanization was introduced into the military and, later, agriculture, the horse began to decline in importance. People still rode horses for enjoyment and show, but a bond had been broken. The animals were no longer necessary for survival.

In recent years, many people have felt the need to escape the noises and smells of the cities and get back in touch with nature. In keeping with this mood, the horse has found favor once more. Riding for fun and sport, classical and sporting dressage, and competitive and recreational driving are gaining in popularity. Of greater significance, though, is the fact that people are again learning to appreciate the intimate and rewarding relationship that is possible between humans and horses and to acknowledge the horse as the personification of vitality, strength, and fluid movement.

Right up to the present day, the stables at Lipica have preserved a precious part of humankind's heritage: the Lipizzaner, one of the purest breeds of horses in the world.

For hundreds of years, Lipica was a somewhat isolated enclave whose chief preoccupation was with breeding. Since the estate's very survival often depended on Masters of the Horse in various courts and armies, its grounds — and its horses — were declared off-limits to the "uninvited," or those who lacked sufficient power and prestige.

At the height of the Austro-Hungarian Empire, Lipica was the exclusive preserve of the Viennese court and the Spanish Riding School. Between World Wars I and II, it was more or less dominated by the military. Only after the second World War did

Lipica begin to look beyond its own stone walls, and in the last twenty years it has opened wide its doors to lovers of horses and equestrian sports from all countries and classes.

The tiny community of Lipica has become internationally famous because of its Lipizzaner horses. The herd has starred in a number of television series and motion pictures, including Walt Disney's 1963 film, "The Miracle of the White Stallions." Lipica horses and carriages have appeared in competition in cities from Aachen to Zagreb and have been sold to buyers in the United States, Italy, Germany, Canada, Argentina, Australia, England, the Netherlands, Sweden, and Switzerland. Troupes of performing Lipizzaners have delighted audiences in many lands.

Lipica's main "industry," however, is tourism. It is a rare traveler who passes through Italy or Slovenia on the way to the Adriatic and does not stop at Lipica to admire the white horses. Thousands of nature lovers visit the estate each year to watch the displays of classical riding. And increasing numbers of equestrian enthusiasts are spending days or even weeks in the company of the horses which were formerly the sole privilege of the Viennese court.

Lipica today is developing in three general directions. First, there is a renewed emphasis on breeding more genuine Lipizzaners with a somewhat wider scope and the capability of learning additional steps designed to meet the demands of sophisticated equestrian sports, particularly dressage and classical riding. Second, there is the need to perfect Lipica's classical school of riding and carriage driving and expand it to include a testing ground for breeding animals. This will not only strengthen the traditional arts of riding, driving, and dressage but also ensure the continuing high quality of the Lipizzaner. And, finally, Lipica is working hard to develop all types of riding and driving facilities for the tourist industry and to organize both greater and smaller exhibitions.

The Lipizzaner and the Art of Riding

Riding has been considered an art since the time of ancient Greece. Xenophon, the eminent Greek historian and statesman, was an accomplished rider whose work, *On Horsemanship (Peri hippikēs)*, described how horses should be handled and revealed his own deep love for the animals by detailing the spiritual relationship possible between horse and rider.

In the centuries that followed the decline of Greece, riding was largely a forgotten art. There were simply too many wars going on and too many nations moving from place to place to allow for its cultivation and perfection. In fact, it was not until the sixteenth century that riding began to regain some of its former stature. Master equestrians appeared in Italy, and the skills they practiced and taught quickly spread to France, Germany, and England.

The Viennese court started emphasizing the art of riding in 1565, and in 1572 the Spanish School of Courtly Riding was founded. At its inception, only Spanish horses were used (hence its name), since they were generally considered the most suitable of the breeds available at the time. In 1735, at the opening of the building that to this day houses the Spanish Riding School in Vienna, fifty-four Lipizzaner stallions took part — evidence that the breeding of horses at Lipica had already reached a high level of perfection. From that time on, the Spanish School has used nothing but Lipizzaners for its classical riding school.

Following the collapse of the Austrian Monarchy, the School established its own breeding stables at Piber near Graz in the southeast corner of Austria. This split the original Lipica herd, and Lipica itself became part of Italy for a time. The estate's connections with the Spanish Riding School were severed, as were its opportunities to continue testing the Lipizzaners' abilities in reputable professional surroundings. The Italian authorities were not particularly interested in the subtleties of classical riding, preferring instead to use the Lipizzaners for elementary riding and carriage driving.

When Lipica became part of Yugoslavia after the second World War, efforts were once again focused on breeding, regenerating the depleted stock of Lipizzaners, and reviving the public interest in the art of riding. In 1952 a group of riders gathered at Lipica under the leadership of the famous trainer Akarov, who was determined to continue the traditions established by the Spanish Riding

Favory Dubovina II was the only stallion left at Lipica in 1965, when most of the horses had to be sold due to financial difficulties.

School. His efforts bore fruit a few short years later, in 1956, when the rider Alfonz Pečovnic rode the mare Thais XI to second place in the middle category at an international dressage event in Vienna.

By 1957 Lipica had twenty-six pureblood stallions and mares ready for training in dressage and carriage driving. In that same year, the management of the riding school was taken over by Milan Belanović, who continued the work begun by Akarov. The results of his expert training were demonstrated at events in Aachen, Verona, and at home in Lipica.

Due to financial difficulties, Belanović's team was soon forced to disband. While a few of the younger riders remained at Lipica, the rest scattered throughout Europe. Even more unfortunate was the fact that most of the trained horses had to be sold. Only one stallion remained, Favory Dubovina II, who was ridden at that time by the young and relatively inexperienced rider Klavdij Maver.

In 1973 the Czechoslovakian trainer Emil Šulgan arrived at Lipica to reopen the riding school. He was able to recruit some of the riders who had worked under Akarov and Belanović, in addition to some new ones. In 1975 Šulgan was replaced by Đorđe Petrović, a retired lieutenant and former leader of the cavalry school team. He began systematically training a group of riders which by then numbered fifteen. At that time, there were more than twenty colts and five stallions in the *haute école,* or advanced dressage school of classical riding. Meanwhile Klavdij Maver was sent to the Spanish Riding School in Vienna. He successfully completed his education

The rider Alfonz Pečovnik and his mare, Thais XI, became a famous team. Here they are shown executing the *levade*.

there in 1977 and was brought back to Lipica to take over Petrović's role as trainer.

By 1979 the Lipica team consisted of fifteen professionally qualified riders and teachers and a group of stallions which had received anywhere from one to four years of training. The Lipica stables were in better shape than they had been in for quite some time, and they continue to improve to this day.

The Schooling of the Lipizzaner

Lipizzaners and their riders are trained according to a demanding program based on the principles and practices of the Spanish Riding School in Vienna. Some of the breeding stallions are also prepared for competitive dressage.

The training begins when three-and-a-half-year-old colts are brought in from the herd to the stable for six to twelve weeks of riderless gentling. During this breaking-in period, the horses are taught absolute obedience and a few preliminary movements. The remainder of the first year is spent getting the horses accustomed to walking, trotting, and finally cantering with their riders.

At the end of the first year, an initial selection is made. Only those horses whose appearance, temperament, and abilities meet rigorous criteria and who have developed along satisfactory lines and worked well with their riders are allowed to continue on to the classical school. Those which do not fulfill these requirements are culled and distributed to tourist stables.

The second year of training is spent in familiarizing the

Pečovnik and Thais XI executing the *piaffe*.

horses with more demanding steps and maneuvers. They are required to master the large and small circles and turns and to learn the correct transitions from the walk to the trot and from the trot to the canter. Toward the end of this year, schooling in straight-line maneuvers — backward and forward steps — is augmented by training in lateral — or side-to-side — movements. Following this, the horses are again assessed according to their obedience, intelligence, development, and appearance. Those which are not judged satisfactory are withdrawn from further schooling.

In the third year of training, the horses pass from the lower school to the high school, or *haute école*. It is now that the steps and maneuvers are perfected and refined. The horses are expected to develop perfect balance as they perform all the figures of classical riding. They must also demonstrate a graceful carriage and pure paces. In addition, they must show an ability to execute collected paces, or a series of different steps in an ordered sequence. Finally, they are taught the extended canter, which aims at the longest possible stride. They are first schooled in the simple change of canter and later in the flying change at two, three, or four strides.

Toward the end of the third year, the horses are trained to perform the half-*pirouette* and lateral movements while cantering. They then learn the *piaffe*, or on-the-spot trot, and the *passage*, or swimming trot (also called the Spanish trot).

In the beginning, the horses perform these complicated exercises "in hand;" in other words, without riders. Only when they thoroughly master the complex rhythms of the *piaffe* and the *passage* and are able to execute them smoothly and confidently do they begin to learn these difficult and elegant figures of the *haute école* under their riders.

Throughout the first three years of training, each horse and its rider must demonstrate perfect harmony with each other. This does not only mean that the horse must be obedient to the rider's every signal, but also that the rider must be responsive to and aware of the horse. In a very real sense, the horse and rider are trained together. They become a team. They develop the kind of spiritual relationship that Xenophon described so many centuries ago.

Pečovnik and Thais XI in an extended trot.

A few figures from the *haute école* (sketches by Ludwig Koch):

the *piaffe*

After the third year of training, the stallions are once again subjected to a series of rigorous tests. Those which pass with flying colors — and which are also perfect in appearance — are sent out to stud for one season. They are accepted into the breeding program on a permanent basis after two or three years if their offspring prove to be acceptable. This ensures that only the best stallions will be used for breeding.

Breeding mares must also undergo a number of tests. When they are three and a half years old, they are removed from the herd and sent to the stables for carriage training. They spend the next six to twelve weeks growing accustomed to free driving and then to driving with a carriage harness. Once they master the lunge line, fillies are harnessed alongside trained mares for a period of time; later, two young fillies are harnessed together. The following six to eight months are devoted to pulling the two-horse carriage; during the final two months of this period, the horses are trained to carry and respond to their riders.

At the end of this time, the first selection is made. Fillies are judged according to the following criteria: appearance, temperament, obedience, their understanding of carriage driving and riding, and the purity and rhythm of their gait. Those which fulfill all the

the *capriole*

the *passage*

requirements of the selection are designated as future breeding mares. Before they are admitted into the breeding program, however, they must spend yet another year learning to pull two- and four-horse carriages.

Stallions which enter the fourth year of classical training are taught to consolidate the figures they learned in the third year. Soon they are able to proceed to the flying change of canter at every stride and execute flawless *pirouettes*, *piaffes*, and *passages*. Once they have demonstrated their skill in these movements, their on-the-ground training is, for all practical purposes, over.

In the second half of the fourth year, the horses begin doing exercises for above-ground training. Those which show an inclination for complex individual figures — including the *levade*, the *courbette*, and the *capriole* — are first taught to do a series of leaps, at first in hand and then beneath their riders. Since these advanced maneuvers are extremely difficult and demanding, only very intelligent and very strong horses are permitted to learn them.

Lipizzaners were being taught the *levade* at Lipica as early as 1956. At that time, the mare Thais XI, who had been trained by Alfonz Pečovnik, was the only horse capable of doing it. In 1977, when Klavdij Maver returned from the Spanish Riding School in

the *ballotade*

the *pirouette*

Vienna, above-ground training was begun with three new stallions. Soon Favory Dubovina II, ridden by Maver, was able to execute this movement as well as Thais XI.

The schooling of a Lipizzaner, then, is a detailed and arduous process that begins when the horse is quite young and proceeds through four full years. Only those horses which are able to pass the rigorous annual tests are permitted to continue on through the most demanding movements of the *haute école* and above-ground classical riding.

Lipica aims at perfection, and its performing stallions are shining examples of perfection itself. The fact that they are able to execute their complicated steps with skill and grace is not due to schooling and breeding alone, however. It also testifies to the boundless love and countless hours their riders and trainers willingly devote to them. And this is the real secret of the Lipizzaners' great success at home and around the world.

the *levade* from the *piaffe*

the *courbette*

the *croupade*

A few pieces from Lipica's collection of carriages:

A training wagon, or *šulvagen*

A break

Lipica Today

Each year, several performances are held at Lipica for the benefit of the public. Any visitor who is fortunate enough to attend one never forgets it.

A typical performance begins with a musical introduction. Then, all at once, a troupe of superb white horses prances into the show arena, their riders dressed in beautiful eighteenth-century costumes. From that point on, the music and the horses move as if they were made for one another. The mood is that of a classical ballet — regal, poetic, and imposing.

The Lipizzaners and their riders also display their expertise outside of Lipica's stone walls. In recent years, they have performed in Austria (in 1971), in Italy (in 1976), and at the Balkan Championships in Romania (in 1977), where they took third place. In Zagreb (in 1978), they placed second and third, and in Athens (in 1979) they placed third.

Today the classical school of riding at Lipica is no longer reserved for professional riders alone. Several years ago, a riding school for guests was established, and now any and all visitors to Lipica who demonstrate an enthusiasm for equestrian sports may enroll in courses there. Special programs are offered for beginners, who are introduced to the basics of riding on quiet, trained Lipizzaners under the guidance of professional teachers. Those guests who have had some riding experience and wish to improve their skills may take more advanced courses. Finally, visitors who desire training in the classical maneuvers of the *haute école* and want to enjoy the true art of riding may participate in a course of classical riding and dressage. They are permitted to handle the best horses and are supervised by the finest and most expert trainers.

Because Lipica aims to please, it also provides facilities for recreational riding and carriage driving. Guests may take leisurely trots through the beautiful parks and woods that surround the estate or go on open-air excursions in comfortable carriages. Many visitors like to top off a day of driving or riding with a picnic or a fox hunt. Pony rides delight even the youngest horse lovers.

In recent years, increasing numbers of people have made their way to Lipica and its superb horses. But Lipica offers more than equestrian sports and shows. Its centuries-old oaks and lindens provide the perfect setting for relaxation and reflection. It is an area rich in history and culture. And its proximity to cities and towns on the Adriatic coast — including Trieste, which is a mere six miles away; Portorož, which is twenty-five miles away; and Opatija, which is thirty-seven miles away — allow vacationers to combine watching or riding the white stallions with bathing in balmy sea waters.

The main reason why so many people travel to Lipica each year, though, is and will remain the Lipizzaners. Some visitors come just to enjoy and applaud the performances. Others come to improve their own riding skills. But still others come to experience the magic that is found only in Lipica — where white manes stream in the crisp Karst wind and the air is sweet with the fragrance of lindens.

A coupé

A Victorian landau

A phaeton

A facsimile of the bill of purchase for the three Spanish *brincos* brought back to Lipica by the Baron Hans Khevenhüller in 1580.

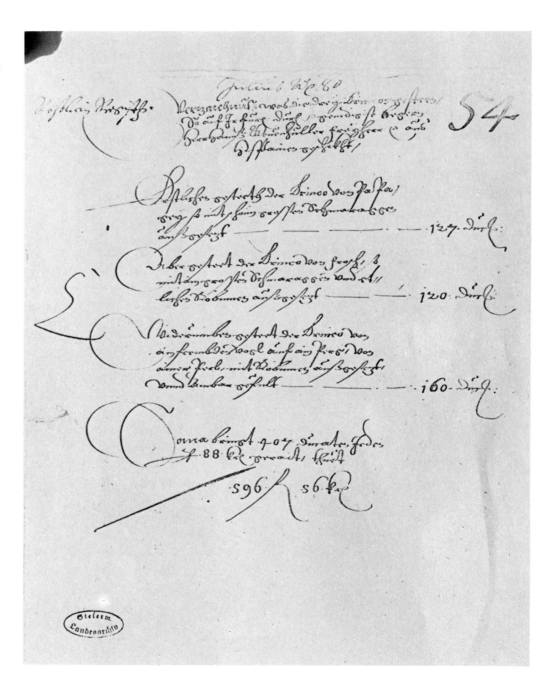

A History of the Lipizzaner

Many people believe that the Lipizzaner was developed solely to satisfy a passing whim of the Austrian court, but this was not the case. To understand the true reasons why the Lipizzaner came into being, one must look back in time more than four hundred years and study the prevailing conditions and needs of that era.

Beginnings and Growth

By the middle of the sixteenth century, Austria was already well on its way to becoming a major world power. It stretched from Switzerland on the west into Hungary on the east, up through Bohemia and Moravia in the north and down through Croatia in the south. Its future looked rosy — and, in fact, was. By the end of the eighteenth century, under the leadership of the Imperial House of Habsburg, it would have more than doubled in size.

To keep up with the demands of running a growing empire in the 1500's, the Austrians had to rely to a great extent upon the horse. Horses moved their massive armies and facilitated economic development. They served politicians, the clergy, merchants, artisans, peasants, and the court in both public and private capacities. It is understandable, then, that the Austrians devoted considerable amounts of time and energy to their care and breeding.

Following the decline of Spanish horse breeding — Spanish horses had long been regarded as the best available in western and central Europe — the Austrians began exploring the possibility of establishing independent local breeding stables. In time, they hoped, they would be able to develop horses of superior strength, intelligence, endurance, and beauty. Those that proved satisfactory would not only serve the highly demanding court but also the postal network. In addition, the best of these would be reserved for the Spanish Riding School, where they would eventually replace Spanish horses altogether.

When Charles, the youngest son of the Emperor Ferdinand I, became Archduke and Regent of what was then called Inner Austria — an area which included Styria, Carinthia, Carniola, Gorizia, Istria, and Trieste — he started taking steps to ensure that an adequate supply of suitable horses would be made available for both civilian and military purposes. In 1576 he paid a visit to Trieste, a city on the Adriatic coast, and spoke with its Bishop Coreta and a man named France Jurko.

Jurko was a good organizer — he was probably the administrator of the Episcopal stables in Trieste at the time — who knew horses and could speak with authority about the soil and climate of the Karst. Both he and Bishop Coreta were well aware of the fact that the Archduke was looking for sites on which to establish Imperial stables, and they had one in mind: a small, isolated estate called Lipica.

Lipica had already been in existence for at least a century. Documents dating from the 1400's make reference to a house in a village called Lipica, in front of which grew a small linden tree, or *lipa*. The house contained a wine shop, and the locals would occasionally drop in for a bit of refreshment *k lipici* — at the linden — and it is from this phrase that Lipica undoubtedly derives its name.

Bishop Coreta and France Jurko encouraged the Archduke to consider purchasing Lipica, and on May 19, 1580 — four years after Charles's initial visit to Trieste — a contract was finally signed. At the same time it was arranged that the Karst horses in the Episcopal stables would be moved to Lipica. Since then, 1580 has traditionally been regarded as the year in which the Lipizzaner breed was founded.

Although the Archduke's decision eventually proved a wise one, his reasons for arriving at it were not exactly clear. True, the Slovene Karst, where Lipica was located, closely resembled Spain and neighboring Italy, where horse breeding was already well established. Perhaps Charles — urged on by Jurko — felt that this was a good omen. Lipica itself was hardly a prize, however. In 1576, the year of the Archduke's visit, it was the half-ruined country seat of the bishops of Trieste. Used as a kind of holiday retreat, it was cared for by poor serfs who struggled to eke out a living on the rocky and infertile land. Even the serfs came and went; sometimes there were as many as seven families living there, at other times as few as two.

In 1559 the Turks under Malkozbeg had carried out a number of predatory raids against the Karst, Lipica included, and the afteraffects — starvation and plague — had been widespread. Although sixteen years had gone by since then, the estate was still in disrepair and showed obvious signs of neglect. So Bishop Coreta may have had more than purely altruistic motives when he first broached the subject of Lipica's sale to Charles.

At any rate, the proper papers were signed, the Bishop was named counselor to the new stables, and France Jurko was given the job of administrator. Immediately afterward, Baron Hans Khevenhüller left for Spain with instructions to purchase Spanish horses that could be crossbred with the solid, strong Karst animals. Khevenhüller brought back three pureblood stallions, or *brincos* — a word which in Spanish means treasures or jewels.

In 1581 six more purebred stallions — including one Andalusion — and twenty-four mares were brought back from Spain and added to the herd at Lipica. By 1585 Jurko was able to report to the Archduke at his seat in Graz that the stables were doing well. The horses were thriving, and the most urgently needed buildings, water cisterns, and other facilities had been built. To protect the horses from attacks by wolves, bears, and lynxes, a stone wall was erected around a courtyard and trained guard dogs were purchased. The dogs proved their worth in 1589, when a veritable invasion of wolves swept across Carniola and the outskirts of Trieste.

In 1594 Lipica bought a neighboring property belonging to a man named Jurij Božič. On this *novi zavodi*, or new farm, were built two open stables and two small houses for taking care of the horses, the ruins of which may still be seen today.

According to a receipt dated 1595, a stable hand was paid a considerable sum of money during that year to take thirty young horses from Lipica to the Archduke in Graz. From this it may be inferred that after only fifteen years of operation the stables were able to send its annual increase of colts and foals to Charles for his personal use and that of the court. Those horses which were not needed at Graz were distributed among other stables for breeding purposes.

In 1596 Ferdinand, Charles's firstborn son, became Archduke and vowed to continue the work begun by his father. With his support, the stables at Lipica were enlarged and special attention was paid to the quality of the new breed being developed there — the Lipizzaner.

On September 7, 1658, the stable master at Lipica received a communication from the newest Archduke containing a lengthy list of instructions. Among other things, it set forth a working order and plan for the stables and requested that even more horses be produced at Lipica than before. At that time the Viennese court was engaging in all sorts of celebrations and ceremonials. It would frequently hold triumphal processions consisting of at least twenty richly ornamented carriages led by a number of riders on horseback. The Lipizzaners, with their proud carriage and high-stepping gait, were already great favorites.

The Influence of the Habsburgs

Under the Emperor Leopold I, who reigned from 1640 to 1705, the stables at Lipica flourished, along with the entire Austrian empire. By the middle of the seventeenth century, the Lipizzaner had reached full flower as a breed. When Prince Eugen of Savoy led his victorious campaign against the Turks, many of his cavalrymen rode the already famous horses.

Meanwhile several members of the nobility were vying with one another as to who could buy the best Lipizzaner from the Imperial stables. To ride one of these fine horses — or harness it to a carriage — brought instant prestige. A number of feudal barons began their own stables and stocked them with Lipizzaners for their knights, who valued them for their speed in battle. Many owners built elaborate stables to house their prizes. Riding schools — which were extremely fashionable at the time — cropped up in countless towns and castles.

During the reign of Joseph I from 1705 through 1711, the herd at Lipica was expanded again, necessitating the building of still more new stables. At the request of Gasparo Nicoletti, Lipica's administrator, a church and a priest's house were also constructed. These were large enough to contain schoolrooms in which the local children received instruction in the Slovene language. By then the once-deserted Lipica was a bustling community of people who were experienced in and skilled at breeding and training horses.

Above the *velbanca* — the stable for purebred stallions — was written an inscription that visitors today can pause to read. It states that during the reigns of the Emperors Leopold I and Joseph I many new buildings were erected at Lipica. This describes only part of the unceasing activity that went on for many years. Older stables were repaired, staff quarters were constructed, pastures were improved, and paths and woods were put in order. Almost all of the buildings of that time are still standing and in use after more than two centuries.

The ascendancy of the Emperor Charles VI, who ruled from 1711 to 1740, brought even greater prosperity to Lipica. Charles VI highly valued the Lipizzaners, and whenever he wished to give some nobleman a special gift he always permitted the lucky recipient to choose one of the horses.

By then the herd had grown so large that the current acreage — which included the *novi zavodi* as well as a mountain pasture at Jurišič that had been rented since 1629 — could not produce enough hay. In 1711 the new administrator, Maks Oblak (later Baron Wolkensperg), rented the former estate of Prince Auersperg at Postojna. Nine years later, Lipica purchased this property and added its extensive meadows and pastures to its holdings. In 1727 a number of new buildings were erected at Postojna, including a large grain storage barn.

Lipica also made use of land at Prestranek on the slopes of the Javornik Mountains. These beautiful pastures, which were owned by the Stična monastery, were in many ways better suited for the horses than those at Postojna. The grazing was excellent, and the Lipizzaner as a breed continued to improve. Several facilities were built at Prestranek, including stables for sick horses and harness animals, a coach house, and storage bins for hay. In 1852 another stable was constructed that was capable of housing 100 foals.

Above the entrance to the *velbanca* are inscribed the names of two emperors who promoted the development of Lipica: Leopold I and Joseph I.

To the right of the entrance to the *velbanca* is a stone slab indicating the year of its completion — 1703.

For many years the Prestranek estate was economically important to Lipica.

Three neighboring mountain estates were also taken over by the burgeoning Lipizzaner industry — Bile, Poček, and Škulje — as well as yet another pasture to the west of Škulje called Stresenca. It became a tradition to take Lipica foals to Stresenca for their first grazing. In its center stood a small castle, or *gospodov Paradajzarjev*, a single-story building which housed two-year-olds. The foals remained at Stresenca from May until early autumn, at which time they were driven to Prestranek. There, when they were three years old, the colts and fillies began their training in the harness. They remained at Prestranek for one more year. Then, following a selection process, the breeding fillies were sent to Lipica, geldings and fillies not suitable for breeding underwent further training before being sent to the court stables in Vienna, and stallions chosen for breeding purposes (approximately ten each year) went to the Spanish Riding School in Vienna. All remaining horses were auctioned off.

Before the completion of the railroad between Vienna and Trieste, young horses bound for Vienna were taken overland on foot by their grooms. More often than not, the reigning Emperor could hardly wait for the horses to arrive before he insisted on seeing them. In connection with this fact, there is an old story that the people of Lipica love to tell. In order to get its point, one must understand that the word "Lipizzaner" was at that time used to refer both to the horses and to the people of the Karst.

It seems that a delegation of Lipizzaners — in this case, the human variety — once traveled to Vienna and sought an audience with the Emperor Franz Joseph I. The chamberlain hurried to the court to announce them. "Take them to the stables and see that they are well fed," the Emperor replied. "I will go and look at them later."

During the early years of the Empress Maria Theresa's rule — she reigned from 1740 to 1780 — both she and her husband, Francis I, took an avid interest in the management of Lipica. Under their guidance, the stables continued to flourish.

Prior to that time, the majority of Lipizzaners had been bred for riding purposes. Now, however, the court decided that it wanted still more horses capable of pulling their heavy carriages. In 1765 Maria Theresa founded a new Imperial stable at Kopčany, and in 1768 she saw to it that twenty of its strongest mares were sent to Lipica for introduction into the Lipizzaner breed.

The Empress demonstrated her support in other ways as well. Documents in the Imperial archives show that in May of 1747 she contributed money for the upkeep of the estate. Twenty-one years later, in 1768, the government — presumably at her suggestion — paid out yet another large sum for the same purpose.

During the Seven Years' War with Prussia, which lasted from 1756 to 1763, the Austrians lost many of their horses, and their armies were forced to buy animals from Russia and Turkey. In response, Maria Theresa issued an instruction stating that horses from the Imperial stables were to be used for breeding in an attempt to alleviate the shortage. One of her sons used Lipizzaners to found new stables for the military at Radautz, Mesöhegyes, Bábolna, and Piber.

By now the reputation of Lipica and its horses was known in many countries around the world. George Hamilton, the well-known English painter of the early eighteenth century, once traveled all the way from Vienna to Lipica to see them. He was so taken by the estate, its surroundings, and its horses that he produced seventy-

A painting of the Lipizzaners by the English artist George Hamilton.

two paintings, some of which can be seen today at the Spanish Riding School and at Schönbrunn.

Lipica has survived a number of close calls during its history, and one of them occurred in 1785. A commission was sent to Ljubljana to ascertain the economic conditions of Slovenia and evaluate Lipica and Prestranek. Afterward, the commission proposed that the Lipica stables be closed and its horses moved to Holič in Galicia. Luckily, Joseph II vetoed this suggestion.

Hard Times: The Wars With France

A far greater threat to the survival of Lipica — and to the Habsburg Empire as a whole — was posed by a series of wars with France which lasted from 1792 to 1815. In 1796 columns of the Austrian army rolled over the Karst and turned Lipica into a military camp. As the French army under the young Napoleon advanced across Carinthia and Styria, it soon became apparent that the horses at Lipica would have to be moved to safety. On March 22 almost three hundred horses were arranged in four columns and started on their way to Székesfehérvár in Hungary.

A large number of mares were with foal at the time, and special care had to be taken of them. By the fourth day of the journey, six mares had delivered, and their foals were gently loaded into the carts and carried the rest of the way. Further along, ten more mares had their foals. A total of sixteen Lipizzaners were born during the arduous eight-day journey, which was made even more difficult by inclement weather, and not one of them or their mothers was lost.

Upon arrival at Székesfehérvár, the herd was split up. Expectant mares and yearlings were sent to Sz. György near Lake Balaton, and the rest of the horses were sent to the estates of Tyhanger and Moor. Purebred stallions and mares which had not yet been mated remained at Székesfehérvár and were mated immediately so as not to waste any time.

In October of 1797, following the Peace of Campo Formio, the foals were returned to Prestranek. Later that same year, the mares and stallions were brought back to Lipica. A group of expectant mares stayed on at Moor.

When Napoleon entered Trieste as a conqueror, the people of that city decided that there was only one gift worth giving him: a Lipizzaner. Local burghers went posthaste to Lipica, chose a stallion, and presented it to the young general. Bonaparte mounted the horse at once, in front of the Brigida Palace, and rode it to Campo Marzio. He left Trieste riding the same stallion.

Meanwhile Lipica was attempting to pull itself together after the devastation caused by the war. Upon their return from Székesfehérvár, the people of Lipica had been astounded to find the estate in ruins. Many buildings had either been severely damaged or destroyed, and the archives — which contained all of the Lipizzaners' pedigree records since the founding of the breed — had disappeared. Whether they had been destroyed by French soldiers when Lipica was evacuated, or packed up for the flight to Hungary and then lost in transit, remains a mystery. Record-keeping was not begun again in earnest until 1815, and at that point it was decided to make duplicates of all documents from then on, with one copy to be retained at Lipica and the other sent to Vienna.

The damaged buildings were repaired as quickly as possible and the ruined ones eventually restored. Lipica's troubles were not over, however. In 1802 a severe earthquake shook the estate, and in one of the stables most of the best horses were killed.

During the same year, the mountain property of Škulje, which up until then had only been leased, was purchased and added to the Lipica holdings.

Following pages: A view of Lipica from the west, painted in 1779.

In 1805 Napoleon declared war on Austria and Russia, and the Lipizzaners were once again forced to flee their home in the Karst. On December 15 the herd began journeying toward Djakovo in the region of Slavonia. Seventeen days later, in early January, they finally arrived. The move was a long and difficult one, complicated by the fact that the roads were often blocked by troops or made impassable by heavy rains. The military had confiscated all fodder and stables along the route, and this meant that the Lipizzaners had to camp in the open and often went hungry.

The horses remained at Djakovo until October of 1806, when they were moved to the estate of Karad in Hungary. The herd had barely settled into renovated stables when fire broke out in the village, spreading so rapidly that the breeding mares were rescued just in time from their burning building.

On April 16, 1807 — after only six months at Karad — the herd was ordered back to Lipica. Three Arabian stallions from Vienna were also brought along.

The return journey was even more difficult than the one in October of the preceding year. The weather was especially bad, and one foal died en route. The retinue arrived in Lipica during early May only to find that the estate had been plundered again.

Under the terms of the Peace Treaty of Schönbrunn, which was signed in 1809, Carinthia and Trieste were made over to France. To avoid letting the herd fall into the hands of the French, the horses were moved a third time. The transport left Lipica with 289 horses and made its way to Pecska in Hungary, arriving after forty-six days on the road. Three mares dropped foals along the way; two of them became ill, and one died.

Pecska lies on the Maros River near Mesöhegyes in the Tisa valley. Its climate, marshy terrain, fodder, and water — everything, in fact — are completely different from the clear air and stony ground at Lipica. During their stay, the horses had to be stabled most of the time and were denied their previous freedom of movement. Over a period of six years, this had a harmful effect on the herd. Their strength ebbed, and many of the mares miscarried.

When the French entered Lipica only to find the stables

empty, they attempted to revive the breed with some of the animals that had been left behind and others which were found in the possession of local residents. They brought in the Arab stallion Vezir from Egypt, a personal favorite of Napoleon. Vezir was later retained for breeding when the herd returned from Hungary.

Meanwhile Napoleon awarded Lipica to Marshal Marmont, Governor of Illyria. When Marmont was transferred to the Spanish Army in 1811, he leased Lipica — along with the properties of Škulje, Bile, Poček, and Ulačna — to a tenant who could have cared less about their history and upkeep. He was interested only in exploiting the estate, and under his direction the forests were barbarously depleted. In memory of this tragic period, a local thoroughfare was renamed Marmont's Road, or *Marmornata cesta*.

Renovation and Renewal

The Lipizzaners eventually returned to Lipica, but it would be years before they regained their former strength. A number of horses had been lost during the ongoing wars with France, and the estate itself was in need of a great deal of work. Old buildings were renovated and new ones constructed; wells were dug, pastures and meadows were put in order, and ravaged forests were replanted. The remaining horses were carefully assessed, and those which did not satisfy rigorous criteria were withdrawn from the breeding stock. The people of Lipica were determined to revive the breed according to the old rules — for them, hardship would not serve as an excuse for relaxing the standards which had been maintained for centuries.

In 1826 the stables at Kopčany — which had been founded by Maria Theresa in 1765 — were disbanded and their horses were moved to Lipica. At this time, the Austrian court was investing large sums of money in reconstructing its empire, and there was not much left over for Lipica's use. Again there was talk of moving the horses elsewhere and closing Lipica once and for all — echoes of the report filed by the 1785 commission — but the Emperor Franz Joseph I would not allow it.

In 1848 cavalry general Karel Grünne, the senior stable

master at the Austrian court, assumed responsibility for Lipica, where he remained until 1875. He tackled his new job with zeal and an aesthetic sense that had been lacking in previous administrators. In 1852 he saw to it that three large stables for breeding were built, along with a covered riding school, a smithy, a stable for sick horses, storage space for fodder, and a veterinarian's residence. The first floor of the old administration building was adapted to include living quarters, and a new well was dug. Special attention was paid to the meadows and pastures at both Lipica and Prestranek, which were fertilized and, in some cases, replenished with soil brought in from elsewhere. Reforestation was also emphasized, and its results can be seen in the well-tended forests and parks of today.

On May 19, 1880, the third century of the foundation of the Lipica stables was celebrated with due pomp and ceremony. To mark the occasion, a book entitled *Das K.K. Hofgestut zu Lipizza (1580-1880)* was published in Vienna.

According to the information available, the estate by that time consisted of five mutually independent units: Lipica, Prestranek, Bile and Poček (considered as one unit), Škulje, and Ulačno. All of the holdings together encompassed nearly 6000 acres and required 195 employees for their upkeep and maintenance. Distributed among the various stables and pastures were 341 horses, including six breeding stallions and eighty-seven breeding mares. Despite its earlier troubles, Lipica was still the best-known horse breeding site in the Austro-Hungarian Monarchy.

Renovation continued during 1893. A new accommodation block was built at Bile, along with new stables and a water cistern. At Lipica, the mangers were lowered and wooden ones removed; since then the horses have fed from the floor. In 1898 the mountain pasture at Ulačno under the Javornik Mountains was traded for approximately twelve hundred acres of pasture land at Gora and Jančerija. In that same year a coach house was built at Lipica, and in the following year a stable for young horses in training was also erected. Named the *abrihtunga,* or preparation stable, it was built during the fiftieth anniversary of the accession of the Emperor Franz Joseph I and was the last stable constructed at Lipica under the Habsburgs.

Several years of relative peace and quiet ensued. The people of Lipica could once again devote themselves to breeding and training their beloved Lipizzaners. Three-year-olds were brought down from Prestranek and schooled in the *abrihtunga*. In the beginning they were harnessed to a teacher, or *šulmajstra* — an older, quieter horse — and taught to lead a school carriage, or *šulvagen*, which had been specially built for that purpose. The carriage had steps on each side for grooms to stand on, who were ready to jump off and grab the bridle if the horses were suddenly frightened or any other difficulty arose. Once two young horses had become accustomed to driving, they were harnessed together and driven through the surrounding countryside.

Special attention was also paid to the mares and the foals. The mares were strictly segregated according to their stables, and each herd was led out to graze by at least three herdsmen, with a fourth bringing up the rear. The foals were allowed to suckle for four or five months, after which time they began cropping hay and oats from their mothers. When the foals were six months old, they were driven to Prestranek on foot. One groom was assigned to every two horses for the duration of the journey.

At Prestranek, the foals were separated according to sex. The colts remained there while the fillies were sent on to Bile. At the age of one year, the colts were sent to Škulje to graze. Two years later, following a selection process, those colts who had been chosen for breeding were sent back to Lipica and trained for twelve months. After this they went on to the Spanish Riding School in Vienna. Horses which were below breeding standards were gelded and left at Prestranek; at the age of four they were sent to Kladruby and trained by the court for carriage driving.

Every October, three-year-old fillies which met certain requirements were sent to Lipica and taught to pull a carriage. Each year, ten to fifteen of these were chosen as breeding mares. In the following March, they were mated. When it was certain that they were carrying foals, they were distributed among the herd mares.

Information available from the years preceding the first World War indicates that in 1894 there were 295 horses in the Lipica stables — forty-six fewer than in 1880. By now, of course, Lipizzaners could be found in many other places throughout Europe. They had been introduced into military stables at Radautz, Mesöhegyes, and Piber. The governments of Hungary, Croatia-Slavonia, and Bosnia-Hercegovina had purchased breeding stallions of their own, and several private breeders were also regular buyers. Not everyone could afford to own a Lipizzaner, however. To give some indication of how much they were worth, a single Lipizzaner was sold in 1894 for a sensational sum of money that was almost seven times the amount normally paid for a carriage horse.

More Hard Times: The World Wars

During World War I, when Italy declared war on Austria, the herd was forced to make its fourth major move. On May 18, 1915, a special messenger arrived from the Emperor with instructions ordering Emil Finger, the stable master, to take the horses immediately to Laxenburg in Austria. Twenty-four hours later, the horses left Sežana and Prestranek in seven railway wagons. The last transport arrived at Laxenburg on May 29.

Breeding stallions, mares, four-year-old fillies, and carriage horses were stabled at Laxenburg, while foals of both sexes were moved to Kladruby in Czechoslovakia. Emil Finger accompanied the horses to Laxenburg and worked tirelessly to protect Lipica's good name and reputation. Only a few overseers, maintenance personnel, and guards stayed on at Lipica and Prestranek.

Although the facilities at both Laxenburg and Kladruby were in excellent condition, the foreign climates did not suit the Karst horses, who suffered in their new locations as they had in Hungary during the early 1800's. They became less sturdy and more susceptible to disease. Poor quality hay did not help matters; thirty-one mares carrying foals soon died. In the following two years, the fertility rate of the mares dropped to ten percent from the normal eighty percent.

Throughout the war, Lipica and its outstations were used as military camps. At the end of the war, when the Italians took over

the estate, General Carlo Petitti, the Governor of the Giulian Province, appointed a commission to go to Vienna and negotiate with the Austrian government for the return of the Lipizzaners. The Austrians were less than eager to comply, since they wished to continue the breed themselves. A lengthy series of negotiations finally resulted in the division of the breeding stock, with the Italian commission taking those horses which would enable them to continue the Lipizzaner breed in its original environment. When the final compromise was reached, Italy received 109 horses and Austria was left with ninety-eight.

    The Austrians moved their herd to Piber near Graz and began breeding their own Lipizzaners. Today these stables supply all of the horses used at the Spanish Riding School. The Czechoslovakians acquired Kladruby with its 137 horses and established their own breed. After more than three centuries, the Imperial stables begun at Lipica by the Habsburgs had ceased to exist.

    Meanwhile the Italians got busy at Lipica. They took possession of the estate on July 16, 1919. The transition was a fairly smooth one, since the Austrian army had already put the living quarters and stables in order. Nor did the Italians have an especially hard task in front of them when it came to breeding, since they had simply taken the horses they wanted from the Austrians. Colonel

Not long before his assassination in Sarajevo in 1914 — an event which precipitated World War I — the Archduke Francis Ferdinand and his wife, Sophie, rode from the castle of Miramare to Trieste in a carriage drawn by Lipizzaners.

Emanuel Bertetti, a veterinarian, was made director and retained that post until July 1, 1923, when the stables were placed under the authority of the Italian Ministry of War.

The Ministry used Lipica for its own breeding and veterinary purposes. They assigned a new director — another veterinarian named Colonel Giovanni Gotti — and gave him precise instructions to increase the number of breeding mares. All of the horses in the stables were closely examined, and those which did not meet certain specifications were sent to military units.

Lipica was now charged with the task of producing larger, stronger horses suited to the needs of farming and the Italian army. An Oriental stallion, Moro, was imported and mated with six mares in an attempt to achieve this goal. The Lipizzaner stallions Maestoso VIII and Conversano Austria, both of which were of exceptional height, were also used extensively for breeding. In 1926 the stallion Favory Noblessa was brought in from Laxenburg. The foals which resulted from the Italians' experiments were fed a new diet intended to increase their size and strength. The results were not especially outstanding.

In 1927 the Arabian stallion Flower was purchased, but neither he nor any of the other Arabian stallions and mares which were mated with Lipizzaners in 1929, 1931, and 1933 produced the desired offspring. In 1935 several original Lipizzaner mares were given to military units as riding horses, while half-Arabian stallions and mares were brought to Lipica to produce more riding horses for the Italian army. The Italians also imported the half-thoroughbred stallions Ugo Capeta and Agrifolia. Their progeny were of good quality but unsuited to the climate and stony terrain of the Karst, and the stallions were removed from the breeding program after three years.

Colonel Giovanni Gotti continued the tradition of sending foals to graze at Prestranek and Poček. In 1930 the Italians started sending their cavalry horses to Prestranek, and by 1939 Prestranek, Bile, and Poček had all been taken over by the army.

By this time the shadow of Adolf Hitler had darkened all of Europe, and World War II was changing the face of the earth.

Lipica may have been a tiny, isolated spot out of the mainstream of events, but its horses were valuable enough to be fought over. In June of 1943, when Mussolini fell from power, the Germans already had their eye on the Karst and its white stallions. As soon as Italy capitulated on September 9, 1943, the Germans occupied Lipica. By October 16 they had loaded the Lipizzaners onto trucks, driven them to the railway station in Sežana, and loaded them for transit to

A scene from the 1963 Walt Disney film, "The Miracle of the White Stallions," depicting the American rescue of the Lipizzaners in 1945.

Hostinec in Sudetenland. All told, they removed 179 horses from the estate, including six stallions, fifty-six breeding mares, and 117 young horses born between 1940 and 1943.

Not surprisingly, the herd was accompanied on its journey by people from Lipica. They had long been devoted to "their" horses and had proved themselves constant and loyal companions in ages past. They had cared for the Lipizzaners during the flights from the French army in the early 1800's and were not about to be separated from them now. So when the trains pulled away from Sežana, there were fourteen Slovenes and several Lipica grooms on board. In addition, there were eight Italians, one of whom was the most recent director of the stables, Dr. Ugo Fasani. Twenty days later — only after making sure that the horses would be well cared for in Hostinec — all but eight Slovenes and one Italian stable boy returned to the Karst.

For the remainder of the war, the Germans used Lipica as a veterinary unit and a base for confiscated horses. Prestranek was organized as a Home Guard defensive post. Following a battle with the Partisans in the autumn of 1943, the property at Škulje was burned to the ground.

Meanwhile Hostinec under the Germans became a sort of camp for horses brought in from all over Europe. As the headquarters of Germany's Supreme Cavalry Command, it soon housed not only the Lipizzaners from Lipica but also those from the stables at Demir Kapija, Piber, Bábolna, Topolčianky, and Janow in Poland, as well as the Arabians from Dušanovo near Skopje. Six additional Lipizzaners were purchased from the Countess Eltz at Vukovar. Other superb thoroughbreds and trotters were brought in from occupied countries — including some 600 horses from the regions of the Don and the Urals. Together they waited out the second World War in safety.

In the spring of 1945, Germany fell. In April of that year, Hitler married his mistress, Eva Braun, and the two of them committed suicide in the ruins of the Chancellery in Berlin. By midnight on May 8, 1945, World War II was officially over in Europe.

When Russian's Red Army advanced toward the Suden-

As in the past, the Lipizzaners spend most of their lives outdoors or in open stables.

tenland in a last-minute sweep, Colonel Podhajski, the director of the Spanish Riding School in Vienna, grew concerned about the Lipizzaners at Hostinec. Without them, the School could not survive. In response to his urgent request, the American units under the command of General Patton's tank division carried out a daring rescue of the white stallions and took them to Schwarzenberg in Bavaria. From there four mares from Lipica, three mares and one stallion from Demir Kapija, five mares and one stallion from Dušanovo, and the entire herd from Piber were taken to Mansbach in Bavaria. The remaining Lipica horses were moved to Scharding in Austria on May 31, 1945. In November of that same year they were moved again, this time to Ried, where they stayed until December.

Following the war, the Americans gave part of the Lipica herd — and its archives — to the Italians. The rest of the horses stayed in Austria, where they were used to renew the Piber stables. Several stallions and mares found their way to the United States and private stables there.

Six Lipizzaner stallion lines:
Siglavy Gaeta I

Pluto Dubovina VI

Neapolitano Allegra XXII

The Post-War Years and the Struggle for Survival

Forces of the Yugoslavian Liberation Army occupied Lipica on 45, but April 29, 19were forced to leave it on June 12 under the terms of an agreement with the Allies. It was then taken over by English and American troops, who turned it into a transport and tank base. They built barracks, cook houses, and storage buildings and adapted existing structures to suit their needs. The covered riding school became a bar, and the *velbanca* — the stable for the purebred horses — became a movie theater. Large numbers of young trees were cut down and used for fuel.

In March, 1947, the English and American forces slowly began taking their leave of Lipica. Slovenia became part of Yugoslavia under an international agreement, and the Yugoslavian authorities began demanding the return of the Lipica horses that had been removed during the war. But it was already too late. The Italians would not relinquish the horses the Americans had given them, while the Spanish Riding School's stables at Piber were not about to give up the horses they had managed to acquire.

So the Yugoslavs were forced to take however many horses they could get — and this turned out to be a pitifully small number. When the dust finally settled, the once large and proud Lipica stables had a total of eleven horses to call their own.

These were brought back from Mansbach in Bavaria, and although the new herd was small it was luckily of high quality. The horses which were returned to Yugoslavia included the older stallion Siglavy Slatina II; the brood mares Capra, Dubovina, and Sistina; a younger stallion named Favory Sistina; the mares Capriola, Canissa, Slavina, and Steaka; the younger mare Dubovina II; and the colt Neapolitano Capra, which died a few days after arrival at Lipica. Several months later, thirteen horses were brought in from the former Demir Kapija stables, including the stallion Maestoso Slavinia VI, the mares Allegra XI, Batosta XII, Gaetana II, and Santa IV; the younger fillies Barbana, Ira, and Thais; and two fillies from the former stables at Dušanovo near Skopje, Thais II and Blanka. Also during that same year, the young foals Maestoso Allegra XV and Favory Kadina XXIII were given to the state stables at Ponoviče near Litija.

By the end of 1947, there were twenty-three breeding mares at Lipica — a reasonable number with which to begin restoring the herd. Under Yugoslavia's new government, the Ministry for Agriculture and Forestry was charged with responsibility for Lipica and its holdings at Prestranek, Bile, Poček, and Škulje. They renamed it the Lipica Federal Agricultural Estate and established temporary headquarters at Prestranek.

Lipica had suffered greatly during the war, as had all of Yugoslavia. The majority of its buildings had either been ruined, burned, or plundered. It would take a lot of money, materials, and labor to restore the estate and put it in working order once again. Not surprisingly, the people of Lipica applied themselves to the task with the same zeal they had shown for hundreds of years; by the end of 1948, most of the reconstruction work at both Lipica and Škulje had been completed.

On May 12, 1948, eleven more mares arrived from the former Demir Kapija stables — Allegra IX, Alga, Batosta VIII, Batosta XVII, Gaetana I, Gaeta, Santa V, Santa VII, Santa IX, Santa XI, and Santa XV. In May, 1949, the Federal Ministry of Agriculture and Forestry awarded Lipica another fifty-four horses, most of which came originally from Demir Kapija. Now there was a solid base at Lipica for the renewal of the Lipizzaner breed. In fact, Lipica's new herd was almost as large as the old one had been under the Italians. Of course, many of the horses were not first class, and good breeding mares and stallions were in short supply, but at least there were several pure-blooded Lipizzaners available.

The estate of Lipica was the only federal agricultural and forestry outpost in Slovenia, and in January of 1950 it came under the administration of the People's Republic. As it turned out, that was not entirely for the best. The government soon decided that Lipica would have to hand over thirty-three of its best horses to the stables at Kutjevo in Croatia for the purpose of establishing an elite center there for the breeding of Lipizzaners. In September, 1949, fifteen broad mares, eleven fillies, and seven yearlings were taken

Conversano Trompetta XII

Favory Allegra XXII

Maestoso Allegra XXII

Six Lipizzaner mare lines:
Dubovina XVIII

Canissa XIV

Capriola XIII

away, and the People's Republic of Slovenia was left with a severely depleted herd.

It was also determined that the other estates belonging to Lipica would be reallocated. Bile and Poček — where the foals had formerly been grazed — were no longer made available, and their loss was deeply felt. The freedom of movement which the young horses had had on these mountain pastures had contributed to the endurance of the breed, and now this was somewhat curtailed.

Not prone to giving up easily, Lipica began replacing its lost stock in 1950, when it bought ten mares and one foal from a horse breeding cooperative at Pleternica near Kutjevo. These horses were not pure-blooded Lipizzaners, however, so their purchase made little sense in terms of breeding. Recognizing that Lipica needed help if it was going to survive, the Djakova stables loaned it the three-year-old stallion Pluto Slava XVIII. In 1951 Lipica began buying breeding stallions from stables in Croatia, Bosnia, and Serbia, and in 1952 it acquired eleven mares from Kutjevo.

In November, 1952, a commission of horse-breeding experts examined and assessed the herd at Lipica and rejected seventy of its horses. Most of these had been bred on farms and were either of mixed blood or Lipizzaner-Arab crossbreeds.

It was extremely difficult to acquire good breeding mares, but the problem of getting suitable stallions caused even greater headaches for the people of Lipica. Out of the three stallions which had been returned from Mansbach in 1947, only one, Siglavy Slatina II, came from the original herd, and he was already seventeen years old. The younger stallions — Favory Sistina, Maestoso Allegra XI, and Neapolitano Batosta XV — began breeding late, in 1949. In that same year, the Federal Ministry for Agriculture and Forestry gave Lipica two more Lipizzaner stallions, Siglavy Santa II and Neapolitano Mara VIII.

Between 1948 and 1955, there were thirteen breeding stallions at Lipica: Pluto Slava XVIII and Pluto Gratiosa IV; Conversano Santa XI; Maestoso Slavonia VI and Maestoso Allegra XV; Favory Sistina and Favory Kadina XXIII; Neapolitano Montenegra III, Neapolitano Mara VIII, and Neapolitano Batosta XII; Siglavy Slatina

II and Siglavy Allegra XI; and Tulipan Zenta V. Over a period of seven years, these stallions were mated with 350 mares and fillies, and the average fertility rate was seventy-five percent — not bad, considering that fodder was in short supply. During this time, 117 colts and 103 fillies were born at Lipica. Of these, twenty-eight colts went to breeding stations in Serbia, ten went to Slovenia, seven went to Kutjevo, and fifteen more went to state properties and private owners. Kutjevo bought six fillies, and state properties and private owners from all republics purchased another sixty-two.

When Yugoslavia's national administration was decentralized in 1953, the stables at Lipica came under the authority of the Republic and local councils which, due to the current economic conditions, were understandably not particularly concerned about Lipica's welfare. Development came to a temporary halt. Figures from 1963 indicate that at that time the herd consisted of only fifty-nine horses.

In recent years, Lipica has bought and exchanged a number of horses, compiling a herd of first-class stallions experienced in dressage and with proven breeding capabilities. These include three from the Maestoso line, twelve from the Favory line, nine from the Neapolitano line, four from the Conversano line, four from the Pluto line, and five from the Siglavy line. Sixteen different lines of mares are also represented in the Lipica stables.

By mid-1979 the breeding herd at Lipica numbered 173 pure-blooded Lipizzaners. With an additional twelve ponies, forty riding horses, and ten horses of other breeds, Lipica could boast an overall total of 235 horses.

In the space of four centuries, the tiny, isolated estate of Lipica has lost its beloved herd and regained it again and again. It has survived wars, invasions, political changes, storms, wolves, and any number of catastrophes both natural and unnatural. Through it all, the people of the Karst have hung on for dear life to their land and their goal: breeding and nurturing the strong, graceful, intelligent Lipizzaners. And it is due to their efforts and those of other concerned people around the world that the white stallions are at Lipica today for all to admire and enjoy.

Wera VII

Thais XXVI

Samira XII

The Lipizzaner Outside of Lipica

Although Lipica is generally regarded as the home of the Lipizzaner — it was here that the horse was first developed and bred — the white stallions have spread through neighboring countries and even into other continents. Because of their superior intelligence, endurance, gait, and temperament, they have long been sought after by the military, the nobility, and others who could afford them (or, in some cases, simply take them).

Some countries have acquired Lipizzaners in one way or another, held on to them, and continued to breed them over the years, while others have opened stables and later closed them for good. Here, then, is a general overview of the Lipizzaner's status in Europe today.

In Slovenia

Since the Lipizzaner originated in Slovenia, it might be expected that the breed would be firmly entrenched in several places throughout that land. Surprisingly, though, it is not. This may be due to the fact that the Slovenes have generally been contented with knowing that their prize horses are safe at Lipica. Or it may be because the heavy, strong Lipizzaner has not suited their needs as well as lighter horses and trotters. Or it may even be due to an inexplicable lack of interest — no one knows for sure. At any rate, Slovenia has long been famous for breeding its own high-quality horses, of which the Lipizzaner is only one. Its advantageous geographical position at the crossroads of many trade routes and its strong commercial ties with surrounding countries have made it possible for the Slovenes to pick and choose among several breeds.

Lipica has had a powerful influence on Slovene horse breeding, however. It began to exert it in force during the eighteenth century, when the master of the court stables in Vienna allowed Lipizzaner stallions to be crossbred with mares from farms in the Notranjska and Primorska areas. During this time Lipizzaners were also used to upgrade the stock of various military stables.

In general, though, purebred Lipizzaners in Slovenia have largely been confined to the estate of Lipica.

In Croatia

In contrast to the situation that exists in Slovenia, the Lipizzaner is widespread throughout Croatia. In fact, it is one of the most numerous breeds in that land today.

Credit for the introduction of the Lipizzaner into Croatia must be given to Antun Mandić, a former Bishop of Djakovo. In 1806, when the Lipizzaners were forced to flee before the invading French armies under Napoleon and took refuge in Djakovo, Mandić saw them and was very impressed by them. Later, when the horses returned to Lipica, he kept three of the stallions.

At that time Djakovo already had stables of its own. Those had been founded in 1506 to provide suitable horses for the Bishops of Bosnia and Srem and were, in fact, the oldest stables in Europe. For centuries they had been producing Arabian horses of excellent quality. The Bishop realized that the introduction of the Lipizzaners into other existing breeds could only improve them.

In 1849 the Archbishop Josip Juraj Strossmayer purchased seven brood mares from Lipica, and these were crossbred with the horses at Djakovo. After 1870 the emphasis shifted to the breeding of pure-blooded Lipizzaners.

At first the Djakovo stables produced predominantly dark horses and some light ones, named "Strossmayer whites" in honor of the Archbishop. The latter were much sought after and fetched high prices. By 1898 the stables were thriving and had fifty-six brood mares, four breeding stallions, and 109 foals. The Djakovo Lipizzaners were very popular for use in the fields and had a strong influence on horse breeding in Croatia as a whole from that time on.

By the end of World War I, the Djakovo stables had eighty brood mares and a promising future. Their crucial link to Lipica was suddenly severed, however, when the Karst estate fell into the hands of the Italians. In 1943, during the second World War, there were eighty-three horses at Djakovo, but many of these were subsequently destroyed. Following the war, in 1947, Djakovo was hard at work breeding Lipizzaners once again. This task was made somewhat easier by the fact that Djakovo inherited horses from other

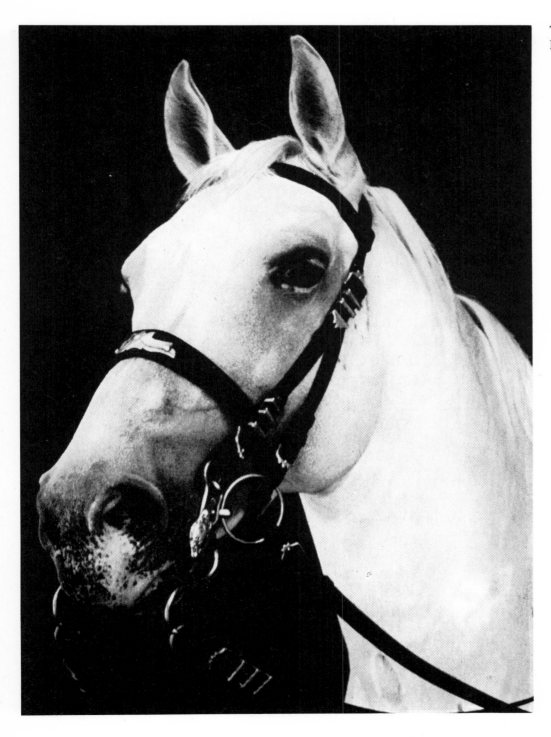

The stallion Favory Allegra XVII came to Lipica from the Djakovo stables in 1974.

stables that were taken over by the military.

In 1959 Djakovo became an official breeding site for Lipizzaners. Ten years later, in 1969, the stables were designated an independent unit of the Agricultural and Industrial Combine of Djakovo. Since 1972 Djakovo has been the home of the Center for Horse Selection of the Socialist Republic of Croatia, or *Centar za selekciju konja SR Hrvatske.*

Currently the breeders at Djakovo are working on producing taller, sturdier horses more suited to farming and other popular markets. As of 1977 they had seven separate sire lines and thirteen mare lines to work with and a total of forty brood mares. The Center is only one of many breeding cooperatives in Croatia, which together own some 400 registered Lipizzaner mares.

Over the years, the Croatians have been able to boast of several other stables devoted to the breeding of Lipizzaners. During the eighteenth century, many large landowners began establishing their own stables. They began by concentrating on Spanish and Italian horses, shifted to Arabians, and finally recognized the value of the Lipizzaners and incorporated them into their herds as well. From there the horses spread into agricultural breeding programs and later to private farms.

The second oldest stables in Croatia were those of Elémer Jankovič at Terezovac. Founded at the beginning of the eighteenth century, they were stocked at first with Hungarian, Arabian, Neopolitan, and later Kladruby horses, most of which were eventually replaced by Lipizzaners. Crossbreeding with old Spanish lines gave rise to the so-called Tulipan Lipizzaners, which were stronger, heavier, and larger than those being bred in Slovenia.

In 1860 a number of mares and stallions were purchased from Lipica and introduced into the Terezovac stables. By 1881 the herd had grown to include fifty-five mares, seven stallions, and numerous foals. Many other well-known stables, including Lipica itself, bought Tulipan foals. By the end of the nineteenth century, the Terezovac stables were among the best and most prolific producers of Lipizzaners in all of Europe. Sadly, the owner at the end of World War I was forced to sell the property and managed to keep only seven mares and stallions. These were soon moved to the Stančić stables.

The stables at Stančić (later called Petrovo) were established in 1919 for the purpose of ensuring a supply of pedigreed breeding stallions for agricultural use. They were stocked with twenty mares bought from private owners, seven from the Kladruby stables, and additional horses from Cabuna as well as Terezovac. Stančić was charged with the responsibility of producing "Croatian-type" Lipizzaners — stronger, sturdier, larger horses than those being bred at Lipica. By 1936 Stančić had five stallions and seventy mares from twenty-two different lines. But in 1937 infectious anemia *(Anemia infectiosa equorum)* broke out among the horses, and by 1938 the stables had been severely depleted. As a result, the breeding program there was discontinued. The same disease caused a great deal of damage throughout all of Croatia, and some mare lines disappeared entirely.

In 1939 new state stables were established at Novigrad na Dobri with thirty-two healthy mares that had survived the infection at Stančić. Another breeding site was founded at Kutjevo following the first World War; by 1936, it had 585 stallions of the Lipizzaner and Nonius breeds.

In 1868 a nobleman named Count Ellz began breeding Nonius horses at his estate at Grabovo. Three years later he opened a new department solely for Lipizzaners, and by 1936 it had twenty-three mares.

Another well-known breeding site was founded at Cabuna in 1945 by Ladislav Janković. It purchased quality stock from several other stables, including Lipica. Mention should also be made here of additional Croatian breeders, including the brothers Tüköry at Daruvar, the Count Ottenfels at Horvatska, Ferdinand Speiser at Orlovnjak, and Stjepan Layer near Lipic. There were at least fifteen more stables in operation at one time or another, including the Terezino polje stables at Virovitica and the Seleš stables near Osijek.

Today the best locally-bred Lipizzaners in Yugoslavia are found in Croatia. The horses are used most often in grain growing areas, indicating that they are excellent farm animals.

In Serbia and Vojvodina

Prior to World War II, breeders in Serbia and Vojvodina used mainly English thoroughbreds and half-bloods in their programs. With Serbia leading the way, though, this situation underwent a series of gradual but important changes.

In 1937 a large group of Lipizzaners consisting of ten mares from Stančić and eight more from breeding associations around Kutjevo arrived at the Ljubičevo stables in Serbia. In 1939 another eight Lipizzaner stallions were sent there from Vukovar and Orlovnjak by order of the Ministry of Agriculture, raising the number of pure-blooded horses at Ljubičevo to twenty-eight.

Ljubičevo was given the task of breeding improved farm animals for use in eastern Serbia, where standards had fallen considerably due to the use of the English horses. Even in the first generation, success was obvious.

Horse-breeding was taken more seriously after that, and this required the purchase of increasing numbers of Lipizzaners. When the Republic of Serbia was divided into three breeding areas, central and eastern Serbia could claim a combined total of thirteen thousand mares, including a strong contingent of Lipizzaners. In fact, the stables at Dobričevo were renewed solely with Lipizzaner horses, including thirty-seven brood mares and four stallions. But in 1954 the breeding site at Dobričevo was disbanded and all breeding stock was sold off.

Following the example set by the Serbians, the breeders of Vojvodina grew increasingly interested in Lipizzaners. In 1929 Branko Ladjević purchased some property at Gladnoš and started a small breeding program there. It ceased functioning before the second World War but was reopened in 1946 under the name Fruška gora. In 1955 a number of Lipizzaners from the Zaravan stables were transferred to Fruška gora, raising its total number of horses to ninety-two brood mares and eighty-one foals. The stables were abandoned in 1959 and their stock was divided up among the remaining stables of the Province of Vojvodina.

In 1946 the Yugoslavian army reestablished the

Karađorđevo stables near Bačka Palanka, which today breed Lipizzaners in addition to Nonius horses, English thoroughbreds, and half-bloods.

In Bosnia and Hercegovina

The Republic of Bosnia and Hercegovina is divided into two horse-breeding areas: northern Bosnia, an area of lowlands and undulating countryside which has been set aside predominantly for the breeding of Lipizzaners; and remaining territories, which tend to be mountainous and have been given over to a local Bosnian breed.

Lipizzaner stallions first appeared in northern Bosnia in 1884, when stables at Sarajevo purchased five of them for breeding purposes. Between 1885 and 1939 the herd grew to a total of 112 Lipizzaners. A number of independent breeders also bought Lipizzaners and introduced them into their own stables.

Lipizzaners were used systematically beginning in 1946, when a new breeding site was developed at Budžak near Banja Luka. By 1949 Budžak had thirty-one Lipizzaner stallions, and by 1962 that number had grown to somewhere between fifty and sixty. When the stables were closed, their stock went to farming cooperatives and private breeders.

It was the establishment of the Vučjak military stables near Prnjavor which guaranteed the continued breeding and distribution of Lipizzaners throughout Bosnia and Hercegovina. Vučjak began operations in 1946 with twelve mares purchased from Lipica. In 1952 more mares of the Djakovo lines were brought in from the Lipica and Kutjevo stables. By 1967 Vučjak had fifty brood mares and two or three stallions.

Interestingly, the military devoted their efforts to breeding dark horses instead of light-colored ones. They achieved this goal by introducing a number of horses from the Tulipan sire line into the stables at Vučjak. They also acquired stallions from the Conversano, Pluto, Siglavy, and Neapolitano lines.

During their first twenty years, the stables at Vučjak produced more than 200 brood mares, and these exerted a strong influence on regional breeding. Traditionally, mare lines are given new names — usually those of rivers in Bosnia and Hercegovina — while the old lineages are also entered into the records. Today the various lines are derived from mares of the Lipica, Djakovo, and Terezovac stables as well as horses purchased from private breeders. All horses carry the "V" brand on their left sides with their registration numbers beneath it. On their right sides they are branded with "BiH" — the initials of the Republic.

In Macedonia

The first Lipizzaners arrived in Macedonia in 1924, when the Italian king, Victor Emmanuel II, presented King Alexander of Yugoslavia with four pure-blooded Lipizzaner mares — Santa, Gaetana, Batosta, and Allegra. Breeding was immediately begun at the newly established stables at Demir Kapija in southern Macedonia on the banks of the River Vardar. Up until the time of the second World War, Demir Kapija produced Lipizzaners of consistently high quality.

Following the war, Macedonia began to take an even greater interest in the breeding of Lipizzaners with a view toward improving local horse breeding. In 1948 a special department for Lipizzaners was established at Vardar and a number of Lipizzaners from Kutjevo were acquired. In 1949 the horses were moved to the village of Gneotin near Bitola; in 1959 the stables there were disbanded.

In Hungary

Long respected as breeders of excellent horses, the Hungarians began producing Lipizzaners when the Emperor Joseph II established new military stables at Mesöhegyes. These benefited from the Napoleonic wars, when the horses at Lipica were evacuated and sent first to Székesfehérvár and later to Mesöhegyes for safety. When the herd eventually returned home to the Karst, a few of its horses were left behind in Hungary.

During the nineteenth century, the stables at Mesöhegyes

progressed steadily, devoting their efforts to the breeding of stallion and mare lines. Although priority was given to the Maestoso stallion line, the new Hungarian-Transylvanian Incitato line was also developed.

After almost sixty years of breeding, someone decided that the Lipizzaners did not really belong on the flat terrain of Mesöhegyes and would do better in the meadows and pastures of Fagaros near the Carpathian mountains. This turned out not to be such a wise decision after all. The horses contracted many diseases at Fagaros, and in 1912 and 1913 the majority of them were moved once again, this time to the old Bábolna stables in northwest Hungary. Those few horses that remained at Fagaros were used by the Romanians in 1919 to found their own Lipizzaner lines.

At Bábolna the Hungarians accomplished a great deal. It was there that they consolidated the Incitato line and began breeding horses of the Tulipan line. Stallions and mares from Bábolna were used in the short-lived Spanish Riding School in Budapest, which was intended to eventually rival the School in Vienna but did not manage to survive the second World War.

In 1788 military stables were established at Radautz in the northern Carpathians. Their task was to raise breeding stallions, including Lipizzaners, for regional use throughout Bukovina and Galicia and also in some areas of present-day Slovenia. A large number of wealthy landowners began stables of their own around this time, the most famous being those of the Count Esterházy at Örylak.

In 1951 all of the approximately 200 Lipizzaners in Hungary were moved to new stables at Silvásvárad under the Bückk hills. This site is still in operation today, since it has proved to be ideally suited for the breeding of Lipizzaners. Once they are weaned, foals are taken up to alpine pastures under Mount Matra, where the fresh air, stony ground, aromatic vegetation, and clear cold water are remarkably similar to those of the Karst. After three years young animals are returned to the stables at Silvásvárad for training in dressage, along with marathon and obstacle driving. Currently Silvásvárad and its surroundings are a stronghold for the breeding of

The stallion Elbedavy, a specimen from an old Hungarian Lipizzaner line.

Five Hungarian Lipizzaner mares parade through Silvásvárad.

Lipizzaners. This is not the only area in Hungary where Lipizzaners are bred, however. The Gyöngyös stables can also boast of fifty stallions.

Lipizzaners are used in agricultural breeding programs throughout Hungary, mostly in the mountain areas of the north. It is estimated that more than a thousand Lipizzaner mares can be found in those regions at the present time. Hungarian horse-breeders have long used Lipizzaner stallions to upgrade substandard lines. Following the second World War, emphasis shifted from producing luxury horses to working stock, and this accounts for the fact that the Hungarian Lipizzaners are the strongest and tallest horses of the breed in the world today.

In Czechoslovakia

The Kladruby stables in Czechoslovakia were founded in 1562 by the Emperor Maximilian II. In 1765 the Empress Maria Theresa established the Imperial Kopčany stables near Holič with Spanish, Lipizzaner, Neapolitan, Danish, and German mares and Lotharingian, Spanish, Kladruby, and Lombard stallions. Kopčany was strengthened by the addition of some Kladruby mares from the Enyed stables in Hungary. In 1815 Kopčany stopped breeding Lipizzaners and began raising English thoroughbred and half-blood horses.

When the horses at Lipica were moved to Laxenburg in 1919, the young fillies and colts were placed at Kladruby. After the collapse of the Austro-Hungarian Monarchy, the Czechoslovakian Republic did not return the foals to Lipica but instead used them to begin their own Lipizzaner lines at Mali Topolčianky near Nitra.

The Mali Topolčianky stables were mainly interested in producing work horses for the hill country of Czechoslovakia. They went along with the desires of the neighboring agricultural breeders and concentrated primarily on breeding dark horses rather than white ones.

In 1944 the Germans gained control over the Mali Topolčianky stables and moved its 600 horses to Hostinec. At the end of the war, only a third of the original herd was returned to the Czechoslovakian Republic, which received thirteen of the original twenty-seven brood mares and not a single stallion. Determined to survive, Mali Topolčianky acquired some foals from Lipica and Piber and a few older horses which had been kept by neighboring farmers. By 1946 the stables had purchased seventy to eighty horses of both sexes. A few of the Lipizzaner foals born at Hostinec in 1943 and 1944 were also brought to Mali Topolčianky, where they were initially used in agricultural breeding programs.

The Lipizzaner breed in Czechoslovakia was restored by three stallions — two whites, Siglavy V and Favory III, and a brown, Pluto III.

As agriculture gradually became more mechanized, the need for work horses waned. In the past decade, however, there has been a revival of interest in sporting horses and this has somewhat benefited the Lipizzaner.

Currently there are nine mare lines in the Topolčianky stables, including those of Theodorasta, Stornella, Presciana, and Deflorata, and an average of between twenty-five to thirty brood mares. Each horse is branded on the left thigh with the letters "CSR" with a capital T superimposed in the center, indicating the lineage; the sire line is shown by a brand on the right side. All fillies are branded on the right side of the neck with their registration numbers.

In Romania

The first Lipizzaner stables in Romania were established at the end of World War I, when the Romanians acquired the foals that the Hungarians had left behind at Fagaros.

The stables, which are named Simbata de Jos, are located in Transylvania at the foot of the Fagaros hills. They currently have some six hundred horses, of which 120 are brood mares. The herd includes six basic Tulipan Lipizzaner lines but no Incitata lines. The fact that Simbata de Jos owns so many Tulipans is evidence of the close ties that exist between the Romanian stables and those at Terezovac in Croatia. Over the years Simbata de Jos has also purchased a number of horses from the Pluto and Favory lines from the same source.

Among the Lipizzaners in Romania, white horses predominate. Approximately seventy percent of the mares are white, twenty percent are brown, ten percent are black, and a few are chestnut. All of the stallions are white with the exception of the chestnut Neapolitano XXI.

In 1970 another breeding site for Lipizzaners was established at Brebena with approximately forty brood mares and six breeding stallions. In the future, the Romanians plan to collect all of the dark Lipizzaners at Brebena and all of the white ones at Simbata de Jos.

Romania's agricultural breeding program includes approximately 8000 Lipizzaners in four separate regions. About fifty percent of the mares are pure-bloods, as are about twenty percent of the stallions.

Apart from the Nonius horse, the Lipizzaner is the most important and popular agricultural breed in Romania, which intends to keep on producing good quality horses for use on its farms.

In Poland

In 1942 the German High Command established a department for Lipizzaners within the well-known Arabian stables at Janow. It had acquired a number of stallions and mares from Stančić and other Yugoslavian stables, and these provided a good start for the breeding program. In 1943 German occupation forces set up more stables for Lipizzaners at Debina, which in both climate and geography is similar to Lipica.

Following the second World War, Debina became part of Russia. From that point on additional breeding stock was brought in from Hungary, Austria, and other places as well. In 1944 Debina was evacuated in the face of the approaching Red Army, and its horses were taken through Hungary to the American zone. From there they were removed to points unknown.

In Italy

During World War II, the herd of Lipizzaners from Lipica was taken to Austria for safekeeping. Following the war, the American authorities in Austria did something which almost everyone but the Italians has wondered about since then: they gave part of Lipica's herd — and its archives — to the Italians.

The 109 horses which the Italians received were used to establish the Monte Maggiore stables at Fara Sabina, a mountainous area some twenty-seven miles from Rome. By 1959 the herd had increased considerably; today, under the supervision of the Ministry of Agriculture, it has approximately thirty-seven pure-blooded Lipizzaner mares.

In Austria

The stables at Piber near Graz were established in 1798 for the Austrian military. They began with extremely mixed stock and did not start breeding Lipizzaners seriously until 1858. At that time they were instructed to produce the Karst horses for agricultural use in Hungary, Croatia, Slavonia, Carniola, and Istra, and purchased several stallions and mares from Lipica.

Between 1860 and 1880 ninety-one breeding stallions were born at Piber — a highly respectable number. Strangely, though, the Viennese authorities were never convinced that Piber was a suitable environment for Lipizzaners, and after eleven years they ordered the stables there to be closed. The stock was distributed among the military stables at Mesöhegyes and Radautz. In 1878 the stables at Piber were reopened and encouraged to produce stronger stallions, for which purpose they imported English and Irish horses.

After the first World War, Piber gave away the horses

The stallion Aly executes a successful *croupade* at the Spanish Riding School.

they had been using up until that time and decided to focus primarily on raising Lipizzaners for the Spanish Riding School. First, though, they had to acquire some good breeding stock — and this was not as easy as it could have been.

The Lipica herd had been taken to Laxenburg for safekeeping during the war, and they might have gone straight from there to Piber had it not been for the Italians. As part of a post-war agreement, Italy had been given Lipica, and the Italians wanted to begin their own breeding program there — with the original herd, of course. After a series of lengthy negotiations, it was determined that the herd at Laxenburg would be divided, with both Italy and Austria receiving horses from all sire and dam lines. In the end, Austria was awarded ninety-seven Lipizzaners — a sizable enough number with which to begin breeding.

Additional Lipizzaners were obtained from the military stables at Radautz. In 1942 Piber had thirty-five brood mares from fourteen distinct lines. During World War II, the Germans moved the herd to Hostinec. After the war, the American occupation forces gave the horses from Piber back to Austria along with part of the Lipica herd. Today Piber's primary responsibility is to supply the Spanish Riding School with Lipizzaners.

At the Spanish Riding School in Vienna

To those who know and love Lipizzaners, the Spanish Riding School is like a country in and of itself — steeped in tradition, redolent with history, and utterly grand. Within its walls, it seems as if time stands still. There are no mundane problems, no everyday concerns — there are only the magnificent white stallions, executing their graceful steps beneath their costumed riders.

In the beginning, of course, the Spanish Riding School used Spanish horses. Slowly but surely, these were augmented and eventually supplanted by Lipizzaners — at first from Lipica and then, after 1916, from the stables at Piber.

The School makes the rules, sets the standards, and establishes the criteria which the breeders at Piber must follow. Over

Lipizzaners and their riders pose in the beautiful arena at the Spanish Riding School.

the years, the School's insistence on quality and perfection has paid off, and today many people believe that the finest Lipizzaners in the world are found here. The traveler who wants to admire the white horses on their native ground goes to Lipica, but the one who wishes to see them performing in all their courtly elegance visits Vienna as well.

The origins of the Spanish Riding School go back to the sixteenth century. At that time the School was very informal and consisted only of an open space for riding (where the court library now stands). As the Austrians became more serious about training horses, they began to realize that it would be far easier to school them indoors where exercises could proceed regardless of the weather or the season.

The building that still houses the Spanish School of Courtly Riding was erected between 1729 and 1735 under the auspices of the Emperor Charles VI. An imposing structure whose interior and exterior alike are considered works of art, the School is the most lavishly appointed arena of its kind in all of Europe. It was opened in 1735 with a splendid display of horseback riding in which fifty-four Lipizzaners took part. Needless to say, this was a great triumph for the Lipica stables, which by then had been in existence for only 150 years. From that time on more and more Lipizzaners from the Karst were used in the *haute école* until they eventually replaced Spanish horses altogether.

The Spanish Riding School soon became a cultural center of major proportions. Under the reign of the Empress Maria Theresa, it was the scene of countless festivities. The Empress herself once led a *quadrille* there in the ladies' carousel. In 1848 the first session of the Viennese parliament was convened inside its walls. Later the School was used purely for classical riding and for teaching young aristocrats.

The School was also responsible for examining young stallions from Lipica and assessing their bone structure, joints, muscles, temperament, and character. For this reason four-year-old stallions traveled from Lipica to Vienna each year, and it was only after they passed the School's rigorous tests that they were considered suitable for breeding.

While the Viennese court used Lipizzaners primarily for carriage and coach driving and for the royal horse guards, the Spanish Riding School encouraged them to develop the full range of their abilities. Recognizing the breed's potential, the School trained Lipizzaners in classical steps and dressage, including the most difficult above-ground figures.

Today the Spanish Riding School is famous throughout the world, and nowhere else has the art of riding achieved such heights of perfection.

The Development and Breeding of the Lipizzaner

*T*he magnificent Lipizzaner of today is the result of centuries of careful breeding. Originally derived from the Slovene Karst horse, it was strengthened and improved over time by the addition of other select lines.

To begin to understand how and why the Lipizzaner came into being, it is necessary to look both at the Karst horse on which it was based and other breeds which influenced its development.

The Karst Horse

Long ago, the peoples who lived along the Adriatic coast in Slovenia were far better developed both culturally and economically than those who inhabited the inland regions. This was due to the fact that they enjoyed, by way of the sea, contact with some of the most sophisticated civilizations of the world. They traded with them frequently, acquiring in the meantime a new set of values — and a new interest in horse breeding. Thus, while the peoples inland concentrated on raising cattle, those who lived by the sea began setting their sights on raising the finest possible horses.

The precise origins of the Karst horse are unknown, although it is highly likely that they go back to ancient times. Roman historians made reference to the extremely strong and durable horses of the Karst. Near the Timav river — which arises beside Ilirska Bistrica, disappears into the Škocjan Caves, and comes to light again some 200 yards before it flows into the Adriatic — there is believed to have been a temple beneath the cliffs which was dedicated to the Thracian god Diomedes, patron of horses. The temple was surrounded by a thick grove in which the ancestors of the Karst horses were supposedly bred.

The Romans were well aware that the peoples of the Karst had marvelously fast horses, and they soon began systematically crossbreeding them with other lines within their vast empire. Before long horses with Karst blood in them were appearing in Roman armies throughout Europe. Because of their speed, they were also used in chariot races and other sporting events.

During the Middle Ages, and particularly in the sixteenth

century, Karst horses were ridden by knights on campaigns and in tournaments. They became not only highly sought after but quite expensive. Even the smaller animals were valued as pack animals. In 1594 a Karst horse usually sold for two to three times the price of an ordinary animal.

In his book, *The Glory of the Duchy of Cariola,* J. W. Valvasor wrote the following in praise of these sturdy animals and their native land:

> The Karst is very stony, yet there grows amongst the rocks the finest and most succulent grass which serves for grazing. Here also are bred the very best horses, which are called Karst horses....[These] horses are famous throughout Europe, especially for their endurance, longevity, and worth as work and riding horses. This is because from their youth they are accustomed to grazing amongst rocks and stones.

In 1662 Stubenberg echoed these sentiments when he wrote:

> The Karst horse is well-known for its strength and exceptionally long life, being capable of working even for thirty years. The Emperor, and later the princes, had [stables], the most famous of which [were] the Edlinger (at Prestranek). These [stables] bred the very valuable and powerful Karst horses.

Other famous stables of the time included those owned by Prince Auersperg at Postojna and those managed by the Bishops of Trieste.

Karst horses were well-balanced, lean, tough, and of noble build and gait, but they were somewhat slow to mature due to the lack of fodder and the fact that they were often used before they were fully grown. According to records kept by an early Bishop of Djakovo, they reached only about five feet in height and were predominantly white in color.

Karst horses were bred and cared for according to entirely natural methods. This meant that during both winter and summer they were out in the open. When there was not enough pasture, they were given hay. In especially bad weather, they were sheltered beneath slab stone roofs. A remnant of one such roof can be seen to the right of Lipica today as one approaches it from the north.

Karst horses were used by local peoples to carry goods to ports or transport items brought by ships to Trieste and Venice. Merchants also used them in their far-ranging caravans. It is interesting to note that even in their early days Karst horses covered a lot of ground. They brought wine back in sheepskins from Trieste, Gorizia, and Vipava and carried salt, oil, grain, linen, mercury, iron, steel, and other goods from faraway places. Trade routes took them to Ljubljana, Graz, Vienna, Salzburg, and elsewhere. They were sold in Italy and other countries throughout Europe. Until the middle of the nineteenth century, a fair of Karst horses was held annually on June 24 — St. John the Baptist's Day — at Štivan near Prestranek, and people came from all over to buy and sell.

In 1580 the Archduke Charles bought the estate of Lipica from the Bishop of Trieste and founded the stables there. As part of the purchase agreement, the Karst horses currently in the Episcopal stables were brought to Lipica.

These, then, were the first "Lipizzaners" — strong, sturdy Karst horses accustomed to the rocky ground of their native land and able to thrive in its harsh climate. They formed the basis for the breeding program that followed.

The Spanish Horse

As soon as the Archduke finalized the arrangements with the Bishop of Trieste, he sent the Baron Hans Khevenhüller posthaste to Spain with instructions to bring back some Spanish horses for use at Lipica. Khevenhüller obliged by acquiring three stallions, or *brincos*. During the following year the Baron purchased six more Spanish stallions, including one Andalusian. The latter proved to be especially significant. It became the progenitor of the "ermines" — the white horses which were later prized by the Viennese court and today symbolize the entire Lipizzaner breed to many people around the world.

As did most other horse-lovers of the sixteenth century, the Archduke Charles subscribed to the Spanish ideal of horse-breeding. At this time it was widely believed that Spanish horses were the best available in Europe, and in many respects this was true. Spanish horses were highly valued, much sought after, and very fashionable. Even as early as the Middle Ages they were profoundly influencing the development of nearly all European horse breeds.

Like the Lipizzaner, the Spanish horse did not merely spring up out of the ground. It, too, had its origins in other lines. During the eighth century, when the Moors overran the Iberian Peninsula, they brought with them horses from northern Africa in addition to a number of Arabian stallions. These, when introduced into existing Spanish stock, had a positive effect. During the almost 800 years of Moorish rule, horse-breeding in Spain reached heights previously unknown in Europe. When the Moors were finally driven out of Spain, horse-breeding there began to decline in quality. With the coming of gunpowder to Europe in the sixteenth century, both heavy armor and heavy horses suddenly became obsolete, and breeders began developing lighter and faster horses.

Nevertheless, many European countries attempted to preserve the noble Spanish lines, mainly because they wanted to ensure steady supplies of good horses for their royal courts. Italy in particular had considerable success in this area during the beginning of the sixteenth century, at which time it established two great breeding centers: one at Naples, and the other at Polesina in the north. Spanish horses were brought in to both places and crossbred with indigenous lines.

Spanish horses became widespread throughout Europe. They were used in the finest stables, and it is easy to understand why the Archduke Charles would have wanted to make sure that some of their bloodlines were introduced into the new stables at Lipica. By doing this, he instituted a tradition of bringing Spanish horses to Lipica that would continue right up until the beginning of the eighteenth century. From the outset, though, it was understood that they would be used only to improve the Karst horses, never to replace them. For the Archduke really believed in the tenacious Slovene breed. For many years, in fact, Lipica was called the Royal Imperial Court *Karst* Stud, and all horses bred there were called *Karst* horses or *Karst*-Lipizzaner horses *(Karster Rasse-Lipizzaner Zucht)*.

Many writers in the past have forgotten the Archduke's true purpose in establishing the Lipica stables. They have focused only on the Spanish horses and neglected to give proper credit to the Karst animals. Some of them have even assumed — erroneously — that the Lipizzaners inherited their high-stepping gait from their Spanish forebears while in fact it originated with the horses of the Karst. Spanish horses acquired their gait by means of dressage training; the Karst horses acquired theirs through natural selection. It quite simply developed over time as the horses learned to make their way over the stony terrain and difficult tracks of their homeland.

The Arabian Horse

Although the Spanish horses purchased by the Baron Hans Khevenhüller played an important role in the early development of the Lipizzaner, they were hardly the only horses brought in to improve the Karst breed. Lipica's administrators were attuned to horse-breeding trends throughout Europe and made sure that other animals of good stock — including some from both Germany and Italy — found their way to Slovenia.

Very few foreign breeds had as marked an effect on the Lipizzaner as the Arabian. While other types of horses helped to increase the Lipizzaner's height, they often brought with them characteristics that were not so desirable. Many of their offspring were ultimately rejected as being unsuitable for the aims of the breeding program, and in the final analysis those animals which tended to be used again and again were the ones that most strongly evidenced the traits of the original Karst horse. The Arabian, however, complemented the Karst line rather then interfering with it.

One of the most important Arabians introduced into the Lipica stables was a stallion named Vezir. He belonged to Napoleon and was one of his personal favorites. Between 1807 and 1816 seven other Arabian stallions arrived at Lipica, with Siglavy being the best of these. Tadmor, another fine Arabian, was also brought in. In 1843 two more remarkable horses were added to the Lipica herd, Gazlan

Stallions from the five old Lipizzaner lines: Neapolitano executing a *capriole*

Pluto executing a *piaffe*

Conversano executing a *renvers*

and Saydan; later, Samson, Hadudi, and the beautiful Ben Azet were purchased.

Arabian mares were often acquired for the purpose of introducing fresh blood into the Lipizzaner lines. In 1826 a large number of mares from Kopčany in Czechoslovakia were brought to Lipica for breeding. In 1856 the Emperor Franz Joseph sent a special commission to Syria, Palestine, and the Arabian desert to buy more of these superb horses. The commission brought back sixteen stallions, fifty mares, and fourteen foals; of these, two stallions and sixteen mares with their foals went to Lipica.

For a while Lipica tried breeding pure-blooded Arabians in addition to Lipizzaners. As it turned out, though, the court was not very happy with the results, and it was soon decided that the Arabian stallions would only be crossbred with the Lippizzaner mares. This process on the whole yielded excellent offspring. Austrian breeders of the time generally agreed that while Arabian blood was beneficial, the best horses were those that had a lot of the old Lipizzaner blood in them. This belief has continued to the present day.

The Formation of the Various Lipizzaner Lines

Today there are eight distinct Lipizzaner stallion lines and several mare lines. The stallion lines include the Favory, Maestoso, Conversano, Neapolitano, Pluto, Siglavy, Incitato, and Tulipan. The last two are of less importance than the preceding six.

The Favory and Maestoso lines originated with two great stallions which were brought to Lipica from Kladruby in Bohemia. The Kladruby stables had been established in 1562 with Spanish horses. When they were destroyed by fire in 1756, their entire stock was transferred to Lipica.

Favory was a blue-brown stallion born in 1779; Maestoso was a grey born in 1773. Their offspring were (and are) consistently excellent and have found their way into other fine European stables as well. This has been especially fortunate in the case of the Maestoso line, which died out at Lipica during the early nineteenth century.

Because Maestoso progeny had been purchased over the years by other breeders, the Lipica line was able to be immediately renewed with the acquisition of the stallion Maestoso X from the military stables at Mesöhegyes in Hungary.

The Conversano and Neapolitano lines were founded by Italian horses. The first Conversano was a dark brown born in 1767 in the stables of Count Kaunitz, while the first Neapolitano was a brown born in Italy in 1790.

The Pluto line originated in Denmark. In 1592 the Danish king, Frederik II, had founded new stables at his castle in Frederiksborg. For more than two hundred years afterward some of the most sought-after horses in Europe had been raised here. The Danish horse was the result of crossing several different breeds — mostly Spanish, Italian, and Arabian. The stallion Pluto — a large, handsome grey — was born there in 1765 and later brought to Lipica; four other stallions were eventually purchased from Frederiksborg and introduced into the Lipizzaner breed. Pluto's offspring found great favor in the Austrian court and were often used to pull its heavy coaches.

Maestoso executing a *pesado*

The progenitor of the Siglavy line was an original Arabian stallion of the same name. Born in 1810, he threw superb offspring, and the horses of that line are generally considered to be very good breeding stock.

The Incitato line — also known as the Transylvanian, Hungarian, or Mesöhegyes-Fagaros line — was used for breeding at Lipica between 1852 and 1854. Eleven progeny resulted, all of which were ultimately rejected; this line is still raised in Hungary today.

The Tulipan line is native to Croatia. It goes back to the year 1810, when it was established by Count Janković at his Terezovac stables.

Some old stallion lines have simply died out over the years, including the Montedoro, Lipp, Montebello, Imperatore, Peppoli, Toscanello, Confitero, Danese, Spagniolo, Boemo, Tuperto, Amico, and Le Fancon lines.

From time to time, English thoroughbred and half-blood stallions were also imported into Lipica, although breeding experi-

Favory executing a *piaffe*

ments with them were seldom successful. Some of their progeny were of good quality, being a bit larger than pure-blooded Lipizzaners, but they tended to be physically weaker with a lower and less lively gait. Added to these faults was the fact that the English crossbreeds were no longer true to the Lipizzaner type. The only stallions which left permanent traces were Millord from the Sardinia dam line, Pilgrim from the Englanderia dam line, and Regent from the Spadiglia dam line.

The most difficult task that the early breeders at Lipica were faced with was that of forming the Lipizzaner breed itself. New blood was very carefully introduced so as not to change the essential characteristics set forth as being both necessary and desirable. Occasionally animals had to be brought in from the outside simply to prevent sterility and inbreeding.

Good mare lines have been every bit as important to the development of the Lipizzaner as its stallion lines. The major mare lines at Lipica include some that were developed there and others that originated in Arabia, Croatia, Hungary, Romania, and Czechoslovakia.

Proven lines stem from twenty mares. Of these, four were grey Karst horses. Argentina was born in 1767; at Lipica, mares of this line are called Slava and Adria. Sardinia was born in 1776; at Lipica, mares of this line are called Betalka. Spadiglia was born in 1778; at Lipica, mares of this line are called Monteaura.

Another five were greys from Kladruby. Africa was born in 1764; at Lipica, mares of this line are called Barbara and Batosta. Almerina was born in 1769; at Lipica, mares of this line are called Slavina and Sai. Englanderia was born in 1773; at Lipica, mares of this line are called Allegra. Europa was born in 1774; at Lipica, mares of this line are called Trompeta. Presciana was born in 1782; at Lipica, mares of this line are called Bonadea.

Two of the mares that originated Lipizzaner lines were blacks from Kopčany. Stornella was born in 1748; at Lipica, mares of this line are called Steaka. Famosa was born in 1783.

Deflorata, a grey from Denmark, was used to establish a separate mare line. She was born in 1767; at Lipica, mares of this line are called Capriola and Canissa. Rosza was born in 1886 and came to the Karst from the Aladar Janković stables.

Five more mares which began new lines were Arabians. Kheil il Massaid, a brown, was born in 1841. Mersuha was brought in from the desert, and her birthdate is unknown. Gidrana was born in 1841; at Lipica, mares of this line are named Gaeta and Gaetana. Djebrin, born in 1862, came from the military stables at Radautz; at Lipica, mares of this line are called Dubovina. Mercurio was born in 1883; at Lipica, mares of this line are called Gratiosa.

Theodorasta, born in 1886, was acquired from the stables of the Baron Kapri at Bukovina. At Lipica, mares of this line are called Wera. Radautzerin was purchased from Radautz, but her line originated at Piber in 1915. And, finally, Rebecca began her own mare line at Lipica following World War II. Her dam was an Arabian mare, also named Rebecca, from the Speiser stables; at Lipica, mares of this line are called Thais.

Of these twenty mare lines, fourteen still survive at Lipica today. Lipica's modern breeding program benefits from the availability of so many different stallion and mare lines. This allows horses to be mated frequently without ever having to be too closely bred with blood relatives.

Mares are usually mated between January 15 and June 15, according to a plan which has been carefully mapped out with the assistance of the Veterinary Department of the Biotechnical Faculty at Ljubljana University. The Department also collaborates with Lipica in the selection and reproduction processes. For example, stallions are chosen for breeding in such a way that all of the active lines are represented in the stables at least once over a two- to three-year period.

Modern Lipica is focusing on breeding horses suited to the rigors of classical riding. Only those horses which demonstrate certain specific characteristics and qualities are permitted to breed. This, of course, is in keeping with centuries-old traditions. Long ago, when the Archduke Charles first purchased Lipica, he wanted its stables to produce the best horses in the world — and that is still Lipica's goal today.

Horses born at Lipica are named according to a time-honored system which helps breeders to keep track of the various lines and monitor their development and growth. Colts are given two names — the first being that of the sire, the second being that of the dam. To give an example, Conversano Dubovina is readily identifiable as the offspring of the stallion Conversano and the mare Dubovina.

Fillies are given only one name and, if applicable, a number. The eighth filly in the Bonadea line, for instance, is called Bonadea VIII. (Similarly, if Conversano Dubovina's mother had been Dubovina III, he would be called Conversano Dubovina III.) In addition, each horse is assigned a registration number.

Lipica horses are branded at the age of one year with a capital "L" on the left cheek. This brand means two things: first, that the horse is of the original Lipizzaner breed; and second, that it was born at the estate of Lipica. No other stables in Yugoslavia that breed Lipizzaners use the "L" brand, but its unauthorized use does occur in other countries. All other legitimate stables which breed Lipizzaners brand their horses with their own symbols.

Following the second World War, Lipica began branding its foals twice: first with the "L," and second with the appropriate registration number. This facilitates easy identification of all the horses.

Some other internationally recognized brands for Lipizzaners do exist, but they are not used in Yugoslavia. Breeders in Hungary, Romania, Czechoslovakia, and Austria occasionally use special combination brands that reveal both the sire line and the dam's sire line.

The Lipizzaner's Appearance

Lipizzaners have a truly distinctive appearance. Their bodies are strong, symmetrical, and elegant. They carry themselves with grace and move with ease. The word most often used to describe them is "noble" — and it fits.

Although all Lipizzaners do not look exactly alike —

A perfect specimen of the Lipizzaner breed — beautifully shaped, high-stepping, and noble.

different stables tend to breed them for different purposes, and either a sire line or a mare line can have its own distinguishing characteristics — they do share some common features.

To begin with, the Lipizzaner tends to be a medium-sized horse. Typically, mares measure a little over five feet in height and stallions about five and a half feet. The horse has a rectangular conformation and a strikingly lean, somewhat large head. The face is elongated and the eyes bright and lively with an intelligent expression. The long, muscular neck is set high, which means that the chest is not especially pronounced. The back is long and strong and, at times, curved. The loins are rounded and long with well-pronounced muscles. The full tail is positioned high and carried well. The chest is broad and deep with well-rounded ribs, indicating power and endurance. The shoulders are frequently short and rather steep. Although both the tail and mane are thick, the hair itself is fine and silky.

There is an old saying that a horse is worth as much as its legs and hooves, and this is certainly true for the Lipizzaner. Its legs are short and its stance firm. The upper leg is longer than the foreleg, and this contributes to the breed's high-stepping gait. The cannons are usually of medium length. The joints are wide and distinctive, and the sinews are fine and clearly defined. Pasterns of both the fore and hind legs are short and elastic. Because of the hard Karst ground,

Plenty of clean, fresh air and room to move are two of the main reasons why Lipizzaners are so healthy.

the hooves are extremely tough. They are also small — in comparison with the rest of the horse — and well formed. Both the gait and bearing of the Lipizzaner are energetic and unique. In a typical step, the foreleg is lifted almost to the horizontal and the knee is bent at a right angle — a perfect position for the exercises of the *haute école*.

The beauty of the Lipizzaner is best seen in motion; its charm lies in its harmony of movement. Stepping rhythmically along with its head held high, its neck erect, and a lively expression on its face, a Lipizzaner on the move — whether performing or simply taking a stroll — is a sight that few people who witness it ever forget.

Although Lipizzaners are usually thought of as being pure white, the horses come in other colors as well. In some Yugoslavian breeding programs today, blacks and browns are favored; in fact, some twenty or thirty percent of all Lipizzaners in Yugoslavia at the moment are dark colored. Interestingly, foals are never born white. Instead, they are black, brown, or even mouse-grey at birth. The ones which are destined to be white change color slowly; the entire process may take between six and ten years.

Finally, Lipizzaners mature rather late. For this reason, they are extremely long-lived. Some reach the age of thirty — at which point they are still capable of working.

A well-known authority on horses named C. G. Wrangel once wrote:

> It is not possible to wish for healthier horses than those at Lipica. Local diseases are unknown; bone malformations...are at most a rarity. The mortality rate is extremely low. It is interesting that the sharp *burja* (north wind from the mountains) raises rather than destroys vitality, while the *široko* (south wind from the sea) creates a general drowsiness...The burning sun of high summer and lack of water have no harmful effects, since [the Lipizzaners] are accustomed to such conditions from birth.

Lipica has always taken good care of its horses — which accounts for their appearance and reputation. Even during times when veterinarians were very scarce in Yugoslavia, Lipica had one of its own. In the past, the veterinarian at Lipica held a very important position: he was second only to the court veterinarian in Vienna.

Throughout most of the four centuries of its existence, the Lipizzaner breed has been remarkably strong and healthy. The only thing that has really seemed to affect it over the years has been a change in climate. Having to flee in the face of one invading army after another has, of course, occasioned a change in climate more times than might have been hoped.

All things considered, the Lipica herd has survived most of its periods in exile relatively well. A total of 300 horses managed to flee to Hungary in 1796 without suffering a single loss, in spite of the fact that a number of mares gave birth along the way. In 1805, when the horses were forced to evacuate Lipica for the second time, only one foal was lost in sixteen months. But in 1809, when the horses were forced to move to Hungary once again — and stay there for six years — they suffered greatly. Their powers of resistance quickly diminished, and in 1810 twenty-two mares miscarried. It took ten years for the Lipizzaners to regain their former strength.

During the second World War, when the horses were moved to Laxenburg, the climatic changes proved fatal to some, while the lack of suitable fodder meant that others starved to death. Fertility fell from the former high of eighty percent experienced at Lipica to an all-time low of ten percent in only two years, and thirty-one of the mares died. Following the war, seven colts and five fillies perished as a result of disease and exhaustion.

Today, of course, the Lipizzaners are back where they belong — at home in Lipica. The tender care given them by the people of the estate and the advice and concern offered by professionals around the world have brought the herd to new peaks of health and well-being. The main problem, it seems, has to do with tourists who are not as careful with the horses as they should be. There have been some incidents of genuine physical damage caused by poor riding techniques and even neglect. One would think that people who traveled to Lipica to ride its white horses would have some respect for them, but this, unfortunately, is not always so.

Following pages: A herd of Lipizzaners graze in a Lipica meadow.

A Map of Lipica

A   The planned entrance-exit complex
B   The planned sports grounds complex
C   Buildings for tourists:
    Hotel Maestoso
    Restaurant
    Swimming pool
    Sauna
    Snack bar
    Bowling alley
    Night club
D   The historical center of Lipica:
    1   Castle
    2   *Velbanca*
    3   Castle courtyard
    4   Lodgings
    5   St. Anthony's Church
    6   Club Hotel Lipica
E   Stables and yards for horses:
    7   Clinic for horses
    8   *Na borjaču* (Stable complex)
    9   Barn
    10 *Abrihtunga* (Training stable)
    11 Training stable
        and small riding arena
    12 Indoor riding arena
    13 Entrance to the stables
    14 Riding school
    15 Courtyards for horses
    16 Stadium
F   Meadows and pastures

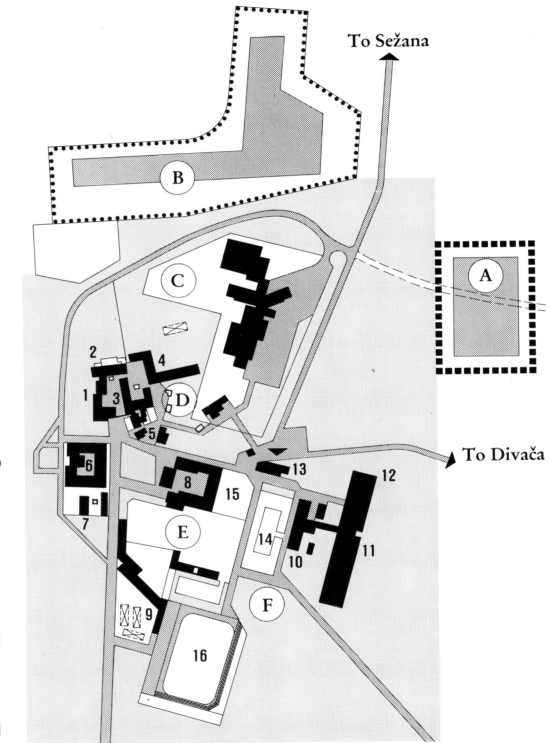

Lipica and Its Surroundings

With its wide avenues, carefully-tended grounds, rolling pastures, fragrant lindens — and, of course, beautiful horses — Lipica is a truly lovely place, as anyone who has visited there knows. And it is lovely in spite of the fact that it has been ravaged, plundered, and gutted more than once by armies and occupation forces which were not overly concerned about its historic worth.

Today travelers come to Lipica from several countries around the world to admire and enjoy it. Many begin their stay with a tour of its buildings.

The Buildings of Lipica

When the estate of Lipica was taken over by the Archduke Charles in 1580, it was in sorry shape. It had been neglected for years, and the only structures it had to speak of were the old Bishop's castle and a little settlement of three serfs' cottages. There were no stables at all and, since the Karst horses from the Episcopal stables were due to arrive whenever Lipica was ready for them, these were urgently needed.

By 1585 France Jurko, the first administrator of the Lipica stables, was able to report to the Archduke in Graz that stables had been built. Today the ruins of what were probably these same stables can be seen to the right of the entrance to the estate as one approaches it from the direction of Sežana.

The central building — the former castle — was once the holiday retreat of the bishops of Trieste. It is situated in the middle of the property on a slight elevation. It was built in three phases, and the extreme left wing — which is the oldest part — is still used today to house the administration.

No one knows precisely when the castle was erected, although it was built in the Renaissance style and it is safe to assume that it dates from this period. To the south is a terrace, laid out as a park; above this is a large paved terrace. Behind the castle to the west is a terraced garden surrounded by a high wall. Facing this garden, on the first floor, were the Archduke's apartments. They were large

The extreme left wing of the castle — the oldest part — is used today for Lipica's administrative offices.

enough to house the Emperor and his retinue, including his chamberlain and assorted courtiers.

The whole castle complex encircles the western part of the inner courtyard, or *hof.* In front of what used to be the priest's house stands a well. At the end of the courtyard is a large stable for breeding stallions. It is called the *velbanca* because of its vaulted entrance.

The *velbanca* is believed to be the second oldest building at Lipica. According to an inscription on a stone slab to the right of the entrance, it dates from 1703. It is a long, spacious, airy enclosure with big windows on the southern side. It is divided along its approximately 131-foot length into ten stalls of equal size. The stalls are enclosed — on the bottom by timbers, and on the top by long

The inside of the castle as seen from the north.

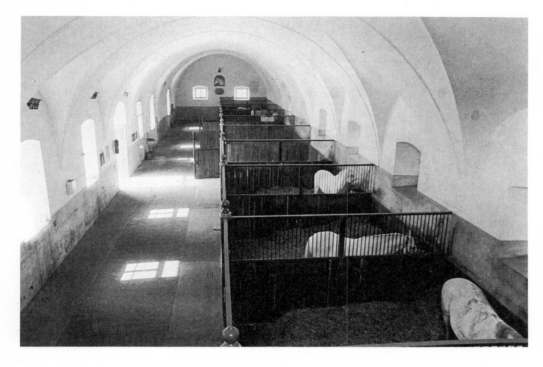

The *velbanca* was built in 1703 and can house up to ten breeding stallions.

iron bars. In the corner of each stall stands a stone manger.

While the *velbanca* originally had a floor made of wooden blocks, this has been replaced by asphalt. A corridor which runs in front of the stalls is paved with pine.

Behind the *velbanca* there was once a pub, in front of which was a terrace with tables and benches. From this comfortable and thirst-quenching position, visitors to Lipica had a marvelous view in any of three directions: north, south, or west.

On the other side of the courtyard, opposite the castle, is a long single-story building. At one time its southern section was used as a stable for official horses, while the manager (or, later, the veterinarian) lived in the northern section. In 1937 the Italians renovated this structure and divided it up into apartments. Following the second World War these were turned into administrative offices and rooms for visitors.

To the west of the *velbanca*, a path leads up to a higher level and a complex of buildings facing a large courtyard. Labeled the "riding school," or *rajtšula*, by the locals, this courtyard is approximately 131 feet long and 65 feet wide. To the south of it is a large old building, rectangular in shape, that was erected in about 1727 in the late Baroque style. Because it was intended for grain storage, it is called the *magacin*. It is characterized by its large cornerstones.

To the right of the *magacin* is a single-story building which runs south-north and then turns eastward. The eastern section is divided into two houses numbered 14 and 15. It is thought that the blacksmith shod horses in house 14 while the veterinarian had his offices in house 15. In the south-north part, called the *kvartirc*, the stable boys and shepherds had their rooms. Adjoining the *kvartirc* and running westward is another single-story building which was probably used as a stable. Nearby was a hayloft, or *fenil*.

In 1938, after the Italians took over Lipica, they turned the former veterinary clinic and smithy into living quarters and added another story.

Near the path leading down from the plateau and the church which stands at its end is a structure which the Italians built in 1933 to serve as their officers' quarters. It has since functioned occasionally as living quarters for the estate's managers and trainers.

On the right side of the path leading to the castle, beside the former priest's house (which today is a restaurant), is a smaller church dedicated to St. Anthony, the patron saint of horses. It is similar in form to other Karst churches of the late Gothic style and is one of the oldest buildings at Lipica.

In the years following the departure of the French forces in 1815, Lipica underwent a period of intense development during which a number of new buildings were constructed. First was a great new stable for mares; this was followed by a covered riding school. Soon three more rectangular stables were erected to the southeast of the castle.

In its day, this stable complex — called *Na borjaču* by the natives — was the largest uniformly designed building at Lipica. In the past, each stable served a separate purpose: one was set aside for foals to use between weaning and training; the second was reserved for mares in foal, those who had just delivered, and those who were nursing their young; and the third sheltered young fillies. Today

The *Na borjaču* stable complex surrounds a central courtyard. To the right is the old riding school.

mares and foals are allowed to move about freely in the first stable. Large windows keep it well-lit during the day, and in the middle of each wall is a huge door that is always left open. A wooden bar is all that prevents the horses from wandering outside. In the middle of the stable complex, which forms a "U"-shape, is a courtyard.

Opposite the second stable and parallel with it is the covered riding school. Inside it is approximately 75 feet long, 41 feet wide, and 13 feet high. The floor is covered with sawdust and the walls are paneled in wood. Large mirrors are hung on the walls to enable riders to monitor their posture.

On the outside wall of the northern side is a plaque which bears the following inscription:

Franciscus Josephus I
Imperator Austriae
MDCCCLII

Or, to translate: "Franz Joseph I, The Emperor of Austria, 1852." At each end of the riding school are wrought iron gates.

Opposite the riding school, across the road near the little park, is a bell-shaped well inscribed with the year 1836.

Some 100 yards away from the entrance to the castle courtyard is a long building which at present houses the veterinary clinic. A stone plaque over the door is inscribed with the year 1769. The building is enclosed on the east and the west by a high wall, and

In the yards by the stables, the Lipizzaners are in daily contact with the stony Karst soil.

to the south is another structure which was formerly used as a stable but now has been converted to a smithy and carpentry shop. To the east of the clinic is a single-story building which was used as a stable as late as the mid-nineteenth century. In 1927 the Italians had it rebuilt, added a floor, and divided it up into apartments.

To the left of the avenue leading to Bazovica is a large partially-enclosed shed which formerly housed the smithy and the carpentry shop; the old blacksmith's forge still stands under the open roof. On the eastern side of this building, behind the present-day smithy and carpentry shop, is a large walled courtyard for foals.

The last stable built at Lipica was constructed by the Austrians in 1898 on the occasion of the fiftieth year of the Emperor Franz Joseph I's reign. It is somewhat removed from the old center of the estate and runs in a southeastern direction. Since it was intended for horses in training, it was called the *abrihtunga*, or preparation stable. It is rectangular and faces the castle. On each end is a room for tackle, and to the center is an area in which fodder is prepared. There are stalls for the horses and some loose-boxes. Today one side of the stable houses fillies while the other has been set aside for tourists. Behind the *abrihtunga* is a large barn, and to the left of the barn is a new stable with stalls for thirteen ponies.

When the Italians occupied Lipica, they set up an open riding school where horses could be trained in good weather. Organized competitions were held here until 1977, when a great new outdoor stadium was built for dressage and carriage driving events.

In 1971 a modern hotel — named Maestoso, after one of Lipica's most famous stallions — was built to accommodate the growing numbers of tourists. The hotel has a covered swimming pool, and to its left are a large open-air pub and terrace.

The Countryside

Lipica is situated on a green oasis in the middle of the Karst, an area known for its natural beauty.

The Karst covers some 300 square miles of land between the Gulf of Trieste, the Vipava Valley, the Pivka River, and the Brkini and Slivnik mountains. Lipica is located at the northernmost part of the Dinar Range in a part of the Karst that is commonly called the "Karst proper," or *Kras* by the natives — a word that is derived from the Latin *Carsium* or *Carstum*.

The region around Lipica is actually the floor of a dry valley which was formed in early times by the Reka River. Over the centuries, the river eroded away the surface of the land and carved

Just beyond the yard for mares and foals is Lipica's new stadium.

out a number of valleys, underground caves, and steep gorges.

The edge of the Karst plateau is so high above Trieste that the village of Opčine is more than a thousand feet above sea level.

Even the ancient Romans were interested in the natural phenomena of the Karst. Virgil and Livy both wrote about them. By the end of the Middle Ages and the beginning of modern times, almost every description of Europe published contained some mention of the Karst's wonders. Of special interest to many people was the disappearance of the Reka River and its reappearance as the Timav.

The Karst's most interesting feature, though, is its multitude of underground caves. More than a thousand have been discovered, and there are twenty-nine in the Lipica region alone, making it one of the richest cave areas in Slovenia.

North of Lokev, not far from the road leading to Sežana, is the Vilenica cave. Prior to the discovery of the Postojna and Škocjan caves, it was thought to be the largest underground cavern in the world. It certainly remains one of the most famous. According to legend, fairies once lived and danced there; even today the fine trail of fog which frequently veils its entrance seems to lend truth to this tale. The cave is nearly 2300 feet long, and at its lowest point it is over 600 feet below the ground. Its many chambers include a "dance hall" in which dances are actually held — perhaps in honor of the fairies.

While Vilenica is still a favorite with tourists, the Škocjan caves are the ones that people most often visit. These are not only a spelunker's dream but are also interesting from an archaeological perspective. When they were first discovered, they contained artifacts from the bronze and iron ages in addition to works dating from the time of the Roman Empire. Tourists began to flock to the Škocjan caves as early as the middle of the nineteenth century. They are fascinating both because of their magnificent stalagmites and stalactites and because of their large water caves.

West of Divača is the Kačna cave, which begins with a funnel-shaped hole that opens wide about 650 feet below the surface. Four long passages lead away from its floor, one of which was discovered in recent years. Nearby are the Košava and Žibernova caves.

The Tomažkova cave, or Crystal cave, is found in the vicinity of Sežana. Near the road between Sežana and Bazovica is the Brunlova cave, which is known for its lovely natural bridges and rock formations. The entrance to the Škamplova cave is on the same road.

At Lipica itself is a large, round, funnel-shaped swallow-hole that contains the one and only spring in the district. Known as *Fontana* — the fountain — it is surrounded by vegetation and fine oak trees.

Another larger swallow-hole is found in the Krkavec region. Huge chestnut trees fill its floor with cool shade, and its sides are covered with oaks, firs, pines, and shrubs. It is a remarkably peaceful place — so much so that a certain cavalry general once requested that he be sent there in the hope of finding a cure for whatever was ailing him at the time. When he in fact did recover, he had a stone altar dedicated to the Virgin Mary erected on the spot.

On the common to the northwest of Lipica are two mushroom-shaped stone tables which are the result of erosion.

Mountain-climbers often take the opportunity to scale Mount Vremščica, which offers a breathtaking view of the countryside. From Mount Slavik one can see the gulfs of Koper and Trieste, the Brkini, the Podgrad hills, and the summit of Čičariji.

The Towns and Villages

Many visitors to Lipica make it a point to visit several of the historic towns which surround it. A general favorite is Sežana, with its three well-tended parks and botanical gardens. Other parks are found at Štanjel, Komen, and Cirje.

There are a number of interesting buildings scattered throughout the Karst. Some date from Roman times, others from the late Gothic and Renaissance periods. Karst people have long been famous for their stone carving and have left behind them many intriguing works.

Located about five miles from Lipica is the village of Lokev, which takes its name from the large trough where animals are watered. Until the first half of the eighteenth century, the main road from the interior of Trieste went through Lokev. Many of the grooms at Lipica still come from this village.

The most imposing building in Lokev is the *tabor*, or stronghold — a tall, cylindrical structure standing at the center of the village. It was built as a defense against the Turks and dates from 1485. It was later used for grain storage. Today it contains apartments. On an outcropping high above Lokev there once stood an older stronghold built at the beginning of the Middle Ages.

Near the *tabor* stands the church of St. Michael, which was built in 1118 by the Knights Templar. The original church was destroyed, rebuilt in 1613, and enlarged in 1628. It is chiefly known for its frescoes, which were painted in 1942 and 1943 by the academic painter Tone Kralj. They depict the part played by the village of Lokev during the second World War. Behind the church is an old linden tree — a trademark of most Karst towns and villages — and nearby is a chapel to Mary built in 1423.

The Illyrian castles, whose massive walls dominate the hilltops throughout the Karst, are even older than the strongholds. The remains of one such castle may be seen on the top of Klemenka near Lokev, and nearby, at Milišče, other traces can be found. On the top of Stari Tabor, above the Vilenica cave, are the ruins of another ancient castle. At one time there were almost ninety castles in the Karst.

The Estate

Lipica itself is situated in the middle of a green oasis surrounded by typical Karst countryside. It is shaded by centuries-old oak trees and sits on a plateau between Sežana, Lokev, and Bazovica, northwest of Trieste and about three miles south of Sežana in one direction and the Adriatic coast in the other. Located about 1300 feet above sea level, it has a slightly undulating landscape that descends rather rapidly toward the west.

The estate is in much better condition than other parts of the Karst largely because it has been taken care of for so many years. It is surrounded by a stone wall and consists of approximately 770 acres of pastureland, meadows, and woods. The topsoil is of varying depth and richness. On elevated, more exposed ground it is brown, while in the hollows it is loamy.

Leading away from the estate are roads to Trieste, Lokev, and Sežana. In the south, cutting through the grounds, is another road running east-west which is used for carriage driving and riding. A road that comes from the direction of Lokev and runs straight toward the castle was built by the French and is called the Marmont Road.

When the Lipica stables were first established by the Archduke Charles, the meadows of the estate were not able to supply enough hay for the horses. Since the soil was too shallow to plow, no grain could be grown. As a result, the food for the horses had to be brought in at first from more fertile areas. In the meantime, the people of Lipica worked hard to put the meadows in order and make them fruitful. They began by clearing trees and uprooting shrubs, destroying protruding rocks, and removing stones. The rock fragments and stones were used to build the wall around the estate; some of the stones were also used as gravel to surface the roads.

Today Lipica produces most of the food it needs for its horses. The pastures are subdivided into plots and used in rotation, and the meadows are divided into western and eastern sections. Oaks, elms, maples, and plane trees grow in the western sections and are pruned back to make hay cutting easier.

Prior to World War I, Lipica and its pastures and meadows were well cared for. The estate was capable of producing more than sixty-one tons of good quality hay each year. During the years under the Italians, the meadows were well fertilized and, for the most part, hay production remained ample, but many of the stands of oak in the pastures were severely depleted. When English and American troops occupied the area after World War II, both the meadows and the pastures were sorely neglected. The more aggressive native trees and shrubs — especially the black elm, black ash, hawthorn, and blackberry — began to invade areas which had been cultivated for several centuries. A number of the oaks decayed and died, while shrubs grew in greater profusion than ever before.

The forests at Lipica were systematically cultivated as early as the sixteenth century. In the second half of the nineteenth century, the estate was fortunate to acquire General Karel Grünne as its administrator. He had a true feel for his work and did a great deal to improve the forests. The famous Montpelier maples date from this time.

Lipica is known for its gracious avenues, most of which were planted during the last century. Today their trees range between 100 and 160 years in age. Especially notable are the plane trees on the Bazovica road. These were planted according to an old custom. In the past, a groom who was about to make his first trip to Vienna had the honor of planting three new trees in the avenues. As a result, the trees are grouped in threes along the road, and each group is separated from the next.

Lipica's avenues and parks were designed to complement its buildings and roads. Two of the finest parks were the direct result of General Grünne's efforts. The first, known to the locals as *parkec* or *vrtec* ("small park" or "small garden"), is located in an area between the castle, the old covered riding school, the church, and the restaurant. The second is at the crossroads to the left of the entrance to the open riding school and the *abrihtunga*.

The green trees and meadows are complemented by splashes of color from the flowers that bloom at different times of the year. In the spring, the striped crocus springs up everywhere; this is followed later by the common peony, the green hellebore, and several other plants that grow in the meadow.

Water has always been scarce in the Karst, and Lipica is no exception. Only one river flows through the entire region: the Reka, which rises up out of the Brkini mountains in the east, goes underground near Škocjan, runs under Lipica, and reemerges between Devin and Monfalcone as the Timav. The Karst has absolutely no surface water, and this has caused continuing problems for Lipica. In 1971 a rather expensive solution was arrived at, and from that time on water has been piped in from the border area near Bazovica.

Lipica has an average annual rainfall of between 56 and 60 inches, each drop of which is considered a gift sent directly from

In the past, ice on the pond had to be broken in order to provide the horses with water. (Both of these photographs were taken in 1963.)

The *Fontana* opposite the riding school carries the date 1836. Behind it is a small park with a few exotic trees.

heaven. For many years, the people of Lipica attempted to drill wells and tap the river they knew ran somewhere beneath their feet, but none of their efforts was successful. Eventually they realized that they would simply have to depend on whatever water fell out of the skies as rain. In 1580 the first *stirne,* or well, was built to collect rainwater. Today carefully constructed wells can be seen near every building. Swallow-holes were artificially enlarged to hold water for the horses and other animals and patched with cement to make them leakproof. A modern pipeline from Bazovica has made the collection of rainwater no longer necessary.

Lipica's climate can be described as being somewhere between the Mediterranean one of the north Adriatic and the alpine one of Slovenia. Although Lipica is only a few miles from the sea, its high altitude makes it a bit brisker than other coastal areas. On the average, though, its temperatures are higher than elsewhere in Slovenia. The sea air brings with it a number of benefits. Saturated with iodine, salt, and ozone, it is very healthy to breathe and has obviously been good for the Lipizzaners over the years. Winters are mild, springs are warm, and autumns last late into the year.

The Karst *burja* — the sharp wind which often blows from the north and northwest — also affects the climate of Lipica. Although it is not particularly pleasant for those who are not accustomed to it, it is both refreshing and healthy. It is said that the people of the Karst and the Lipizzaners themselves derive their strength from the *burja.* And everyone feels out of sorts when the sultry south wind, or *mornik,* blows in from the sea.

This, then, is the ancient estate of Lipica — a land of stones, meadows, forests, blossoms, and fresh breezes; a place where the Lipizzaner has thrived for four centuries and continues to grow even more beautiful with each passing generation. Over the years, many people have read about it and seen photographs of it. But the best way to really learn about Lipica is to experience it. The people of the estate are eager to welcome guests and make them feel at home. And many tourists, travelers, and horse-lovers today are accepting their invitation to visit Lipica and its most famous residents — the Lipizzaners.

The Plates

7

11

2

15

16

27

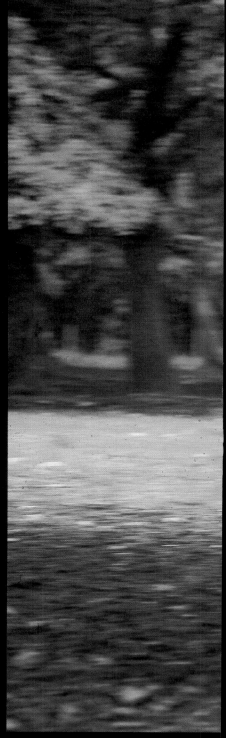

31

32

37

38

41

42

60

61

62

63

64

65

72

73

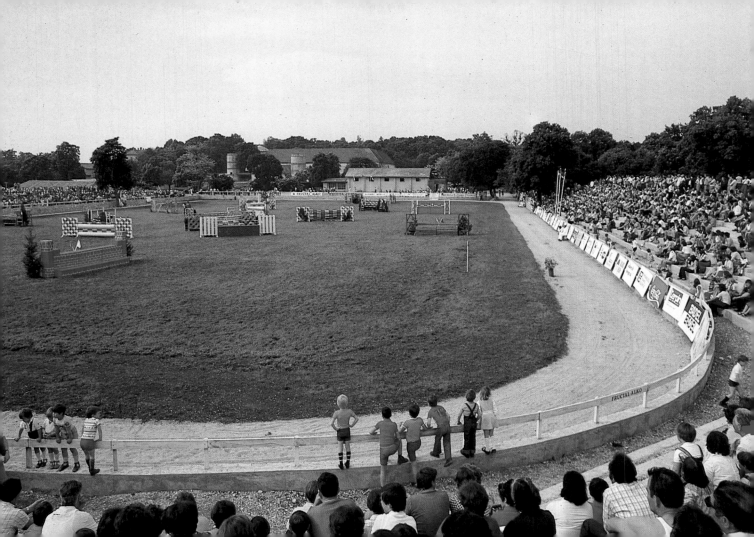

80

81

89

90

91

92

The Plate Captions

1
For four centuries, Lipizzaners have been at home in the rocky Karst.

2
Even the young spend most of their time outdoors.

3
Lipizzaner foals are born dark and do not begin changing color until they reach the age of six.

4
A painting of the Lipizzaners by the English artist George Hamilton.

5
In front of the *Na borjaču* is a wide yard for mares and foals.

6
The horse in the foreground is beginning to change color from black to white. The others show grey spots that will disappear after a few years.

7
The delicate smoke bush, with its colorful blossoms, is found throughout the Karst.

8
Just beyond the yard for mares and foals is Lipica's new stadium.

9
An aerial view of Lipica and the surrounding countryside.

10
A peaceful afternoon in one of Lipica's beautiful valleys.

11
At Lipica, the horses graze for seven or more months each year.

12
Although the Karst grass looks modest, it is very nutritious.

13
One of the many flowers that brighten the Lipica landscape.

14
In the stables, the Lipizzaners are tied up only at mating time.

15
A great event: the birth of a Lipizzaner.

16
The mare's hard work is almost over.

17
As daylight brightens the stable, the mare and her foal get acquainted.

18
With its mother's encouragement, the foal stands up on shaky legs.

19
The mare watches carefully as her foal takes its first tentative steps.

20
Foals suckle for four to five months.

21
Foals are allowed to run in the courtyard between the stables.

22
Foals are expected to remain close to their mothers. If one happens to wander off, it is gently — but firmly — nudged back.

23
As the day comes to an end, the mares take a few more mouthfuls of hay while their foals scamper around the courtyard.

24
Mares and foals often graze together in the carefully tended pastures.

25
A bite of juicy grass, a sip of milk, fresh air, and plenty of room to move — what more could a foal want?

26
A meadow at Lipica.

27
Green grass, green trees, bright sun — it's a perfect day for grazing.

28
The foals have plenty of opportunities for play.

29
One of the mushroom-shaped stone tables along the tourists' riding path.

30
These colts have just completed a day of training — and it feels good to run free!

31
It is important for a Lipizzaner to be shod correctly. This horse is having its hooves rasped.

32
The nails must be set firmly and correctly.

33
Following World War II, Lipica began branding its horses with their registration numbers.

34
The "L" brand on the horse's left cheek means that it is a genuine Lipizzaner born at Lipica.

35
When a foal is three and a half years old, it is separated from the herd and brought to the training stable to begin working on the lunge line.

36
The early stages of the training process can be difficult for both the horse and the trainer.

37
The horse soon learns the proper carriage and gait.

38
After that comes the first encounter with the saddle.

39
The saddlery and tack are always clean and ready for use.

40
It takes time to get accustomed to the saddle.

41
This colt is about to be mounted for the first time.

42
The colt is first trained to carry its rider at a trot.

43
After a training session, the horses are put back with the herd and allowed to run free.

44
Fragrant meadows, full trees, and magnificent horses — this is what the summer visitor to Lipica sees.

45
Oaks are the most common trees at Lipica.

46
Night is falling, and it's time to grab one last bite of grass before heading in.

47
Although the Lipizzaner enjoys running, it is not particularly suited to racing. Its talents lie in the demanding steps of the *haute école*.

48
While Lipizzaners are friendly and sociable animals, this one is feeling shy at the moment — as is indicated by the position of its ears.

49
Even a stallion in love has heavy hooves. During mating time, great care must be taken to keep the stallions from accidentally injuring the mares.

50
There's nothing like a nibble of grass to make the cares of the world disappear.

51
In a moment this herd will turn and gallop away.

52
A bath in the pond offers relief on a hot afternoon.

53
What a life!

54
With their manes and tails flying and their muscles tensed, Lipizzaners at a gallop are poetry in motion.

55
These four youngsters are beginning to change color.

56
Absolute peace and quiet are only part of what Lipica has to offer.

57

58
A stallion learns to execute the *piaffe,* or trot-on-the-spot, between two pillars firmly embedded in the floor of the training stable.

59
A Lipizzaner's natural gait is graceful — and after years of training it's perfect.

60
This horse has been working hard.

61
A Lipizzaner and its rider are a team.

62
This horse has mastered the steps of the *haute école.*

63

64

65

66

67

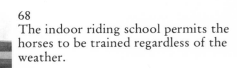
68
The indoor riding school permits the horses to be trained regardless of the weather.

69
Even in the midst of a difficult step, the Lipizzaner keeps its dignified bearing and perfect posture.

70
A performance in the indoor riding school.

71

72

73

74
A performance outdoors.

75

76
This is how Lipica greets its visitors.

77

78
A scene inside the new stadium just before a performance.

79
The stadium is also used for jumping events for horses of other breeds.

80
A visitor mounts a Lipizzaner.

81
All courses at Lipica are taught by experienced trainers.

82
Lipica has a never-ending stream of visitors.

83
The *velbanca* is one of the oldest and most popular buildings at Lipica.

84
Drinking at the trough may not be as romantic as drinking at the pond, but the water is better.

85
Although the herd at Lipica is large, only a few horses are chosen each year for the *haute école*.

86
A fox hunt is about to begin. The leader has attached a fox tail to his sleeve — and it's up to the others to catch him.

87
And the race is on!

88

89
The hunt continues over the vast plains surrounding the estate.

90

91
The red-coated rider is getting close!

92
After the hunt, there's time to let the horses crop a bit of meadow grass before heading for home.

93
The privilege of riding in a carriage drawn by Lipizzaners was once reserved for royalty.

94
Early autumn at Lipica.

95
These horses are hoping for a cool breeze.

96
The forest paths are strewn with leaves — a sure sign that winter is coming.

97
An old oak tree, covered in ivy and reaching toward the blue Karst sky.

98
This wall has been standing for centuries.

99
Tiny Shetland ponies like this one are favorites among visiting children.

100
A colt romping in the midst of wild autumn colors is for many the true symbol of Lipica.

101
The old *Fontana* is a silent reminder of Lipica's early years.

102
Many visitors have rested beneath this tree and its spreading branches.

103
One of the natural wonders of the surrounding countryside.

104
Inside the beautiful Vilenica cave.

105
The village of Lokev — home to many of Lipica's grooms.

106
The village of Štanjel, a monument to Karst architecture.

107
The "Dwarf's House" *(Škrateljnova hiša)* in Divača.

Bibliography

Archivalia of the Country Archives in Graz
Ackerl E., Lehmann A. H.: Die edlen Lipizzaner und die Spanische Reitschule, Wien 1952
Bednarik R. I.: Goriška in tržaška pokrajina v besedi in sliki, Gorica 1932
Benčević Z.: Slavonsko-posavski konj, Stočarstvo 1950, 188-193
Benčević Z.: Lipicanac, Osijek; Djakovo 1957
Benčević Z.: Belić I.: Bijelo biserje Jugoslavije, Zagreb 1968
County records of the Trieste Diocese, Archivio diplomatico, Biblioteca civica
Das K.-u. K. Hofgestüt zu Lipizza, 1580—1880, Wien, 1880
Dossenbach M. in H. Köhler H. J.: Die grossen Gestüte der Welt, Bern 1977
Elitni ergelski centri u Jugoslaviji, Analiza stanja, rada i problemi, Jugoslovenski poljoprivredno-šumarski centar, Beograd 1965, 1966
Ergele u Jugoslaviji, Jugoslovenski poljoprivredno-šumarski centar Beograd 1963, 1964
Fasani U.: Lipizza, paradiso dei cavalli bianchi, Novara 1941
Finger E.: Das ehemalige K.u.K. karster Hofgestüt zu Lipizza 1580—1920, Laxenburg 1930
Frankovič Tone, Lokev, oral information
Gassebner H.: Die Pferdezucht in den im Reichsrathe vertretenen Königreichen und Ländern der Österreichisch-Ungarischen Monarchie, Wien 1896
Giollo R.: I bianchi cavalli di Lipizza, Trieste 1973
Gregorevčič A.: 400 let Lipice in njene zgodovine, typewritten text 1954
Gregorevčič A.: 400 let Lipice in 375 let njene kobilarne, Živinorejec, Ljubljana 1955, 301—311
Hrasnica F.: Uzgojna analiza lipicanske ergele "Vučjak" kod Prnjavora, Veterinaria 1957, 13—33
Ilančić D.: Nekadašnje ergele Slavonije i Srijema, Stočarstvo 1975, 85—110
Jurkovič J.: Stanje u jugoslovenskom konjarstvu i njegova problematika, Zbornik IV. kongresa veterinara i veterinarskih techničara Jugoslavije, Ljubljana 1976, 255—268
Landhauptmanschaft für Krain, Camerale L. (1748—1769)
Lehner H.: Piber, München 1977
L'equille di Lipizza, "Ill Popolo di Trieste" 31. X. 1926
Lisec R., written information, Ljubljana
Mikulec K., Jadras F., Kamenski Dj.: Dugoročni program razvoja konijarstva SR Hrvatske, Stočarstvo 1975, 73—83
Nissen J.: Das Sportpferd, Stuttgart 1963
Ogrizek A., Hrasnica F.: Specialno stočarstvo I., Zagreb 1952
Penko M., Lipica, oral information
Petru S.: Nekaj antičnih zemljevidov in pojmov o naših krajih, Arheološki vestnik XIX — 1968, 3—6
Pferdezucht-Enquete, Wien 1876
Postojnsko okrajno glavarstvo 1889
Program turističnega razvoja Lipice, Biroplanturist, Zagreb 1971
Rapajić N.: Primeri naprednog rada u stočarstvu SRS, "Stočarstvo" 1949
Ravbar M.: Naravna in kulturna dediščina območja Lipice, Zavod SR Slovenije za spomeniško varstvo (typewritten text), Ljubljana 1977
Romano J.: Istorijski osvrt na stanje i razvoj ustanova za uzgoj konja na teritoriji Jugoslavije od XIV. do XX. veka, Stočarstvo 1967, 139—161
Romić S.: Ergela Lipica, velika kulturna vrednota našeg naroda, Stočarski list, Zagreb 1955
Romić S.: O poreklu lipicanskog konja, typewritten text 1978
Romić S.: O uzgoju lipicanca u Hrvatskoj, Veterinarska stanica II, 1971, 77—80
Romić S., written information 1978

Romić S.: Uz 465. obljetnicu ergele Djakovačke, Agronomski glasnik 1972, 645—652
Šček V.: Lokavske starine, typewritten text
Schnelik O.: Das Gestüt der weißen Traumpferde, Die Blauen Hefte 1967
Sila M.: Trst in okolica, Trst 1882
Smerdu R., Ljubljana, oral information 1978
Stefančič A.: Začetek in razvoj veterinarstva na Slovenskem do prve svetovne vojne, Konjereja, Ljubljana 1966
Steinhausz M.: Linije pastuha i rodovi kobila hrvatskog lipicanca, Zagreb 1943
Steinhausz M.: Uzgoj konja, Zagreb 1936
Stipić L.: Bijeli lipicanci Djakova, Djakovo 1975
Stipić L.: O konjarstvu u SR Rumuniji, Veterinarski glasnik 1975, 469—472
Stipić L., written information 1977, 1978
Stopar A., Lokev, oral information 1977, 1978
Valvasor J. W.: Die Ehre des Herzogtums Krain, Ljubljana—Nürnberg 1689
Veble F.: Stanje naše konjereje leta 1934 v številkah, slikah in tabelah, Konjerejec dravske banovine, Ljubljana 1934
Woch Z., Wroclaw, Poland, written information 1978
Wrangel C. G.: Die Rassen des Pferdes, Stuttgart 1909
Wrangel C. G.: Ungarische Pferdezucht in Wort und Bild, Stuttgart 1893
400 godina ergele Djakovo 1506—1977, Djakovo 1977